NUCLEAR TECHNIQUES IN GEOCHEMISTRY AND GEOPHYSICS

Donated by the publisher in the interest of further education in geophysics.

American Geophysical Union Book/Periodical Show

The following States are Members of the International Atomic Energy Agency:

AFGHANISTAN
ALBANIA
ALGERIA
ARGENTINA
AUSTRALIA
AUSTRIA
BANGLADESH
BELGIUM
BOLIVIA
BRAZIL
BULGARIA
BURMA
BYELORUSSIAN SOVIET SOCIALIST REPUBLIC
CAMBODIA
CANADA
CHILE
COLOMBIA
COSTA RICA
CUBA
CYPRUS
CZECHOSLOVAKIA
DEMOCRATIC PEOPLE'S REPUBLIC OF KOREA
DENMARK
DOMINICAN REPUBLIC
ECUADOR
EGYPT
EL SALVADOR
ETHIOPIA
FINLAND
FRANCE
GABON
GERMAN DEMOCRATIC REPUBLIC
GERMANY, FEDERAL REPUBLIC OF
GHANA
GREECE
GUATEMALA
HAITI
HOLY SEE
HUNGARY
ICELAND
INDIA
INDONESIA
IRAN
IRAQ
IRELAND
ISRAEL
ITALY
IVORY COAST
JAMAICA
JAPAN
JORDAN
KENYA
KOREA, REPUBLIC OF
KUWAIT
LEBANON
LIBERIA
LIBYAN ARAB REPUBLIC
LIECHTENSTEIN
LUXEMBOURG
MADAGASCAR
MALAYSIA
MALI
MAURITIUS
MEXICO
MONACO
MONGOLIA
MOROCCO
NETHERLANDS
NEW ZEALAND
NIGER
NIGERIA
NORWAY
PAKISTAN
PANAMA
PARAGUAY
PERU
PHILIPPINES
POLAND
PORTUGAL
REPUBLIC OF SOUTH VIET-NAM
ROMANIA
SAUDI ARABIA
SENEGAL
SIERRA LEONE
SINGAPORE
SOUTH AFRICA
SPAIN
SRI LANKA
SUDAN
SWEDEN
SWITZERLAND
SYRIAN ARAB REPUBLIC
THAILAND
TUNISIA
TURKEY
UGANDA
UKRAINIAN SOVIET SOCIALIST REPUBLIC
UNION OF SOVIET SOCIALIST REPUBLICS
UNITED KINGDOM OF GREAT BRITAIN AND NORTHERN IRELAND
UNITED REPUBLIC OF CAMEROON
UNITED STATES OF AMERICA
URUGUAY
VENEZUELA
YUGOSLAVIA
ZAIRE
ZAMBIA

The Agency's Statute was approved on 23 October 1956 by the Conference on the Statute of the IAEA held at United Nations Headquarters, New York; it entered into force on 29 July 1957. The Headquarters of the Agency are situated in Vienna. Its principal objective is "to accelerate and enlarge the contribution of atomic energy to peace, health and prosperity throughout the world".

Printed by the IAEA in Austria
January 1976

PANEL PROCEEDINGS SERIES

NUCLEAR TECHNIQUES IN GEOCHEMISTRY AND GEOPHYSICS

PROCEEDINGS OF A PANEL ON
NUCLEAR TECHNIQUES IN GEOCHEMISTRY AND GEOPHYSICS
ORGANIZED BY THE
INTERNATIONAL ATOMIC ENERGY AGENCY
AND HELD IN VIENNA, 25 - 29 NOVEMBER 1974

INTERNATIONAL ATOMIC ENERGY AGENCY
VIENNA, 1976

NUCLEAR TECHNIQUES IN GEOCHEMISTRY AND GEOPHYSICS
IAEA, VIENNA, 1976
STI/PUB/425
ISBN 92-0-041076-6

FOREWORD

In the exploration and subsequent recovery of mineral resources, which in certain countries constitute an important part of the national wealth, nuclear techniques play a significant role. They have, for example, been successfully used for a number of years in underground and surface exploration, in laboratory analysis, in process control and in studies carried out at the mine and in the field. Although certain promising techniques for this kind of work are still in the development stage, others have by now become routine procedure in advanced countries.

In 1968 the International Atomic Energy Agency convened a Symposium in Buenos Aires on Nuclear Techniques and Mineral Resources, the proceedings of which were published by the Agency in 1969. This covered a wide range of topics from field experience to the latest developments in instrumentation. Since then the growing interest in the application of these techniques has provided a strong incentive for the development of better instrumentation and for refinements in the techniques themselves, making them in turn more acceptable to industry.

Aware of these advances, the Agency convened a Panel in Cracow in December 1969 on Nuclear Techniques and Mineral Resources in Developing Countries to review the status and to identify those techniques considered most useful to developing countries in the development of their mineral resources.

The principal aim of the Agency Panel on Nuclear Techniques in Geochemistry and Geophysics, held in Vienna from 25 to 29 November 1974, was to evaluate the latest ideas and thinking, and to recommend to the Agency appropriate actions for promoting the most suitable and worthwhile techniques and applications for the Member States. Techniques discussed included those of use in the exploration of oil, gas and minerals as well as in process control. The Panel also considered the relative advantages of nuclear and non-nuclear methods and the important question of standardization.

A summary of the findings of this Panel is presented here, together with individual contributions prepared by Panel Members.

CONTENTS

LOGGING

GEOPHYSICAL WELL LOGGING USING NUCLEAR TECHNIQUES

R.L. CALDWELL, K.P. DESAI, W.R. MILLS, Jr.
Mobil Research and Development Corp.,
Field Research Laboratory,
Dallas, Texas,
United States of America

Abstract

GEOPHYSICAL WELL LOGGING USING NUCLEAR TECHNIQUES.

Nuclear logging of boreholes is in world-wide use for locating and evaluating commercial deposits of hydrocarbons, coal, uranium and certain other useful minerals. Within the petroleum industry the principal uses of nuclear logs are for correlation between wells, for measuring the porosity of rocks, for identifying hydrocarbons and for determining rock type. The long-spaced, dual detector neutron log is especially used to measure porosity and, in combination with a scattered gamma-ray density log to identify gas. Small-diameter borehole accelerators are used as sources of pulsed 14-MeV neutrons to make routinely neutron die-away logs to distinguish hydrocarbons from salt water in cased holes. New accelerator logs based on spectral gamma-ray measurements are being developed to distinguish hydrocarbons from fresh water and to improve lithology determinations. Natural gamma-ray spectral measurements are used to solve correlation problems and to identify uranium-, thorium- and potassium-rich minerals. In coal exploration natural gamma-ray and scattered gamma-ray density logs are used to locate the coal and, in favourable circumstances, to estimate ash content. For oil shale evaluation the density log has been used in small-diameter holes in Colorado oil shale to determine potential oil yield. A new method of uranium detection based on measurement of delayed fission neutrons produced by bombardment of the formations with 14-MeV neutrons from a pulsed accelerator source has recently been described.

I. INTRODUCTION

Geophysical logging for the evaluation of rocks penetrated by a borehole has been developed over the past 45 years into large-scale commercial operations used world-wide in the search for hydrocarbons, coal, uranium and other useful minerals. Table I gives a summary of the principal uses of nuclear logs. This paper covers primarily developments within the United States of America where petroleum industry logging and other services using tools lowered on a cable into the borehole are performed by five major logging service companies, and by many smaller ones. The wireline services of these companies generate revenues of about 400 million dollars yearly. About one-fourth of this income is from nuclear logging which includes measurement of natural gamma rays and the use of neutron and gamma-ray sources to determine physical and chemical properties of rocks and fluids. Gamma-ray and neutron logs are run in practically every new well drilled and also in many old wells to help re-establish commercial production.

The principal uses of nuclear logs in the petroleum industry are for correlation and for measuring the porosity of rocks which serve as reservoirs of oil and gas. The natural gamma-ray log is used for correlation; neutron and scattered gamma-ray density logs are primarily used to

TABLE I. NUCLEAR WELL LOGS

Objective	Logs used
Well-to-well correlation	Gamma ray, neutron
Fluid-filled porosity	Neutron, density
Hydrocarbon identification	
(a) Oil-cased hole	Neutron die-away
(b) Gas-open hole	Neutron, density
Lithology	Neutron, density
	Spectral gamma ray
Coal	
(a) Identification	Gamma ray, density
(b) Ash content	Density
Uranium	Gamma ray, neutron
Oil shale	Density

determine porosity. Since the responses of these two logs to fluid-filled porosity is somewhat different, a combination of these two logs is especially useful for identifying gas and is also effective for determining rock type (lithology).

Pulsed neutron logs are used routinely in cased holes to distinguish hydrocarbons from salt water. New accelerator logs based on spectral gamma-ray measurements are being developed to distinguish hydrocarbons from fresh water and to improve lithology determinations.

Spectral gamma-ray logs of natural radioactivity are run by two service companies. Simultaneous recordings are made of total gamma rays and of the intensity within energy windows selected to be representative of uranium, thorium and potassium. The multi-trace logs have been useful in solving difficult correlation problems and in identifying micaceous formations by the high potassium readings. The gamma-ray detector is sodium iodide.

A recent paper and a US patent describe new methods for uranium detection based on the use of a pulsed source of neutrons. Following a burst of fast neutrons, measurements are made of the time distribution in rock media of either epithermal neutrons or delayed thermal neutrons from fission of uranium. A pulsed 14-MeV neutron source and a ^{3}He neutron detector are used for the measurements.

For logging coal the common procedure is to use natural gamma-ray and scattered gamma-ray density logs to locate the coal. Electrical logs are also commonly used but generally are not as definitive. Under certain conditions the density log can be used to estimate the ash content of coal.

Small-diameter scattered gamma-ray density tools have been used in slim holes in the Colorado oil shale to determine potential oil yield. Calibration of the logs was done in borehole models of known density.

Comparison of log results with assays of cores taken from the holes logged showed good agreement.

High resolution solid-state detectors for spectral gamma-ray analysis are not yet in common logging usage.

II. NEUTRON LOGGING FOR POROSITY

The neutron log is primarily an indicator of fluid in formations and, hence, is a porosity log. The commercial practice today is to run the neutron log simultaneously with the gamma-ray log which identifies shales by their high radioactivity. Formations with low natural radioactivity and low neutron counts are interpreted as porous with the pores filled with hydrogenous fluid. To the steady-state neutron log there is no appreciable difference between oil, fresh water and salt water. Under favourable circumstances the neutron log has been used to locate gas-fluid interfaces because of the lower hydrogen density in the gas phase compared with liquid petroleum. Recently the combination of neutron and density logs has been quite successful in identifying gas in both open and cased holes. Neutron logs have been used for groundwater studies and this use will probably increase in the future [1].

A. Porosity

Neutron logging for porosity determination uses a capsule source of fast neutrons (usually Am-Be or Pu-Be) and a detector of either neutrons or gamma rays. The source generally has an output of about 10^7 n/s. In sedimentary rocks a fast neutron from the source typically travels 0.1 to 0.5 m in slowing down to thermal energy, depending on the hydrogen content of the rock. The average diffusion length of thermal neutrons is only 2 - 5 cm in highly hydrogenous media, but it is considerably greater in dry rock. Thus, in hydrogenous media, the average neutron is captured within a few centimetres of the point where it becomes thermal. The total distance of travel (slowing and diffusing) is an inverse function of hydrogen content. In reservoirs free of clay or shale, the log response is a measure of hydrogen content and, hence, porosity, since hydrogenous fluid fills the pores. Calibration of the log is based on results in test wells constructed of rocks of known porosity and on comparison of logs and core data from field wells.

The log can be run in both open and cased holes — liquid filled or empty. A number of borehole factors influence the open hole measurement: borehole diameter, mud type and weight, temperature, water salinity and the neutron absorption properties of the rock matrix. Standard curves showing the effects of these various factors are used to make suitable corrections. Each logging service company supplies a set of standard calibration curves to users of its equipment and services. In cased holes the accuracy is further diminished by the effects of casing and cement behind the casing.

B. Improved porosity logs

With the development of the high-pressure helium-3 counter as a sensitive neutron detector [2] and the later development of high-output neutron sources, it became possible to make commercial logs based on epithermal neutron detection. Of all the neutron methods tried commercially, epithermal neutron detection provides the simplest determination of formation hydrogen content. Hydrogen plays a very important role in the process of slowing down fast neutrons to epithermal energies; and therefore, the slowing-down length and the flux of epithermal neutrons is determined primarily by hydrogen content. Absorption effects are relatively unimportant in the process of moderating neutrons from high to epithermal energy levels. By detecting epithermal neutrons, the perturbing influences of thermal neutron absorption by the rock matrix and salts in the water are minimized compared with the results obtained with previous tools using thermal neutron or gamma-ray detectors.

SNP log

The advantages of epithermal neutron detection were put to use in the "sidewall epithermal neutron porosity" tool (SNP) for logging uncased holes [3]. In this tool, a Pu-Be or Am-Be source and an epithermal neutron detector are mounted in a skid that is pressed firmly against the borehole wall. The directional detection system emphasizes neutrons arriving primarily from the formation rather than those coming from the borehole fluid, thereby reducing the influences of both borehole and mud properties. A computer panel in the logging truck is used to make minor corrections for mud weight, salinity, borehole diameter and temperature. The log is scaled directly in "porosity index %", based on calibration for a limestone matrix. For more accurate porosity determination one must apply a correction for mud cake thickness. For other rock types slightly different porosity index % scales must be used to get true porosity. The SNP log, although a great improvement over the thermal neutron log, has serious limitations. The tool is not operational in cased holes and the porosity values can be very optimistic in the presence of thick mud cake or in holes of irregular diameter where it is difficult to press the skid firmly against the borehole wall.

CNL log

To overcome the shortcomings of the SNP log a dual-detector, long-spaced neutron log was suggested in 1967 [4]. Two bare helium-3 counters for thermal neutron detection are located in the tool at long spacings from a high output neutron source (4×10^7 n/s). On a relative basis, the strong attenuator for both epithermal and thermal neutrons is the liquid-filled borehole, and the weak attenuator is the formation. For increased sensitivity to the properties of the formation, the detectors should be positioned as far from the source as possible, allowing neutrons that travel toward the detectors within the borehole to be attenuated relative to those travelling in the proper direction within the formation. This problem has been studied using one- and two-group neutron diffusion theories. These studies and experimental measurements have shown that the ratio of thermal neutron fluxes at two appropriately large distances along the borehole axis is virtually identical to the epithermal flux ratio and is a measure of a single epithermal parameter of the formation; i.e., the slowing-down length, which we have already seen, is strongly dependent on the hydrogen content. The studies also showed that variations in salinity of the liquid in the formation or in the borehole and changes in borehole size should have relatively little effect on the flux ratio. The ratio measurement should also not be much influenced by the presence of iron casing and cement.

Commercial logging utilizing the dual-detector concept began in 1971 under the name CNL for "compensated neutron log" [5]. Service is now available from two major logging companies. The tool can be run in both open and cased holes — liquid filled or empty. The ratio of the counts at the near and far detectors is calibrated in terms of porosity for certain open and cased hole conditions. For other borehole conditions, correction charts are available. The CNL tool is frequently run simultaneously with the FDC density log. Under this condition the CNL log is automatically corrected for hole-size variations using the caliper signal from the density tool.

C. Gas detection

The running of two logs in combination saves costly rig time and helps to achieve good depth matching of the logs. The combination CNL-FDC tool is especially useful in exploration for gas [6]. The presence of gas in a formation reduces the bulk density, thereby giving a relatively high porosity index on the density log. On the other hand, the presence of gas reduces the number of hydrogen atoms present in the formation compared with the number in the liquid-filled porosity, thereby indicating a relatively lower porosity on the neutron log. In a water-bearing section the two logs read about the same porosity. A typical example is given in Fig.1 where gas is indicated in the zone from 3340 to 3380 ft.

Although the CNL log is an improvement over other types of neutron logs under most circumstances, it has certain weaknesses also. The presence of elements with large thermal neutron absorption cross-sections has little effect on the epithermal neutron log, but has an appreciable effect on the dual-spaced thermal neutron log. This problem has been discussed in detail in explaining the reasons for the relatively high porosity recorded by the CNL log in dolomite [7]. Thus, there is still a need for research on a logging tool to determine formation porosity with a minimum of effects from all other parameters.

D. Future developments in porosity logging

Probably the most severe interference to porosity determination by the CNL arises from the presence of elements that are strong thermal neutron absorbers. Such elements are boron, lithium, rare earths, and possibly chlorine. As shown by Allen et al. [4], if the thermal neutron detectors in a CNL-type tool are not spaced sufficiently far from the neutron source, an undesirably large response to strong absorbers can exist. One way in which this problem can be overcome, while still retaining the desirable features of a dual-detector system, is through the use of epithermal rather than thermal neutron detectors. For given source strength and source-detector spacings, the expected epithermal count-rate will be lower than the thermal count-rate for two reasons: (1) the epithermal flux will be less than the thermal flux, and (2) epithermal detectors are generally less efficient than thermal detectors. The design of a dual epithermal

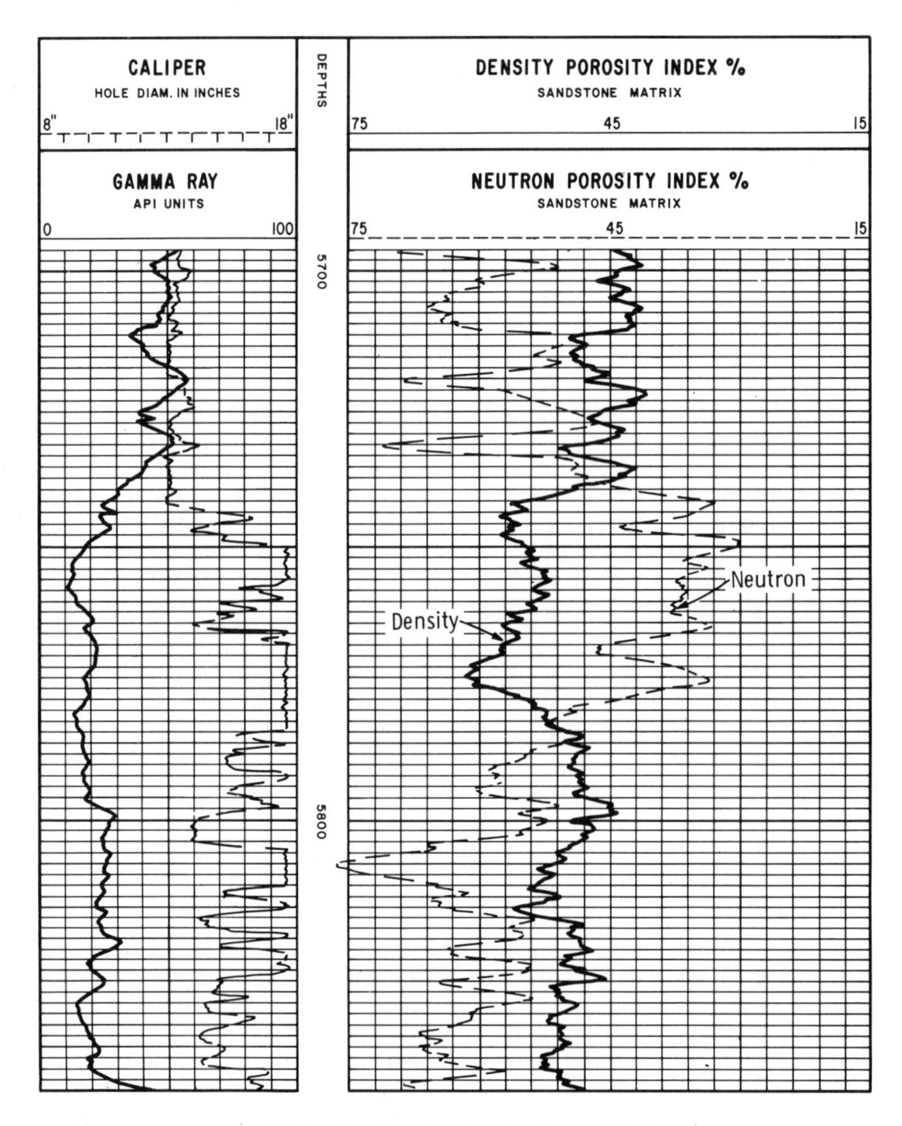

FIG. 1. Density and neutron logs for gas detection.

neutron detector logging tool will involve a compromise between satisfactory counting rates and desired response characteristics.

A measurement of hydrogen content based on the die-away of epithermal neutrons from a pulsed source has been proposed and studied for planetary surface analysis [8, 9]. This method has been shown to have better sensitivity to hydrogen and less sensitivity to non-hydrogenous constituents than steady-state methods. Application of the epithermal die-away technique to borehole geometry will be confronted primarily with instrumental problems, although these do not appear to be more severe than those already overcome in carbon/oxygen logging (see Section III-B).

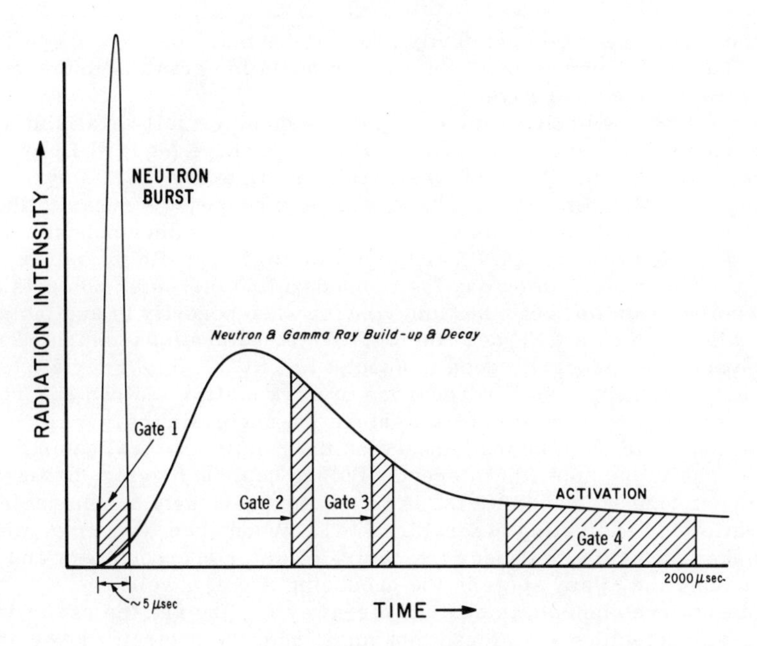

FIG. 2. Pulsed neutron logging.

III. PULSED NEUTRON LOGGING

After more than a decade of research practical logging tools using miniature accelerators to produce pulsed neutron sources have been built. Borehole accelerators of diameter as small as 4.3 cm are available as sources of pulsed 14-MeV neutrons to make a variety of logs using neutron and gamma-ray detectors. The time rate of decay (τ) of neutrons in the formations is a measure of the thermal neutron absorption cross-section, Σ_a, of the rock and its fluid. Since Σ_a is much larger for salt water than for oil, one can distinguish these two fluids in rock by a continuous logging technique.

New experimental logs using accelerators as pulsed sources of 14-MeV neutrons are based on spectral measurements of the gamma rays produced by various neutron interactions. The spectrum is divided into segments by detection at selected time intervals to measure separately the gamma rays arising from inelastic scattering of fast neutrons, radiative capture of thermal neutrons and neutron activation as shown in Fig. 2. One can thereby make selective logs of the carbon/oxygen ratio to distinguish oil from fresh water and of calcium, silicon and aluminium to identify carbonates, silicates and shales.

A. Neutron die-away log

To measure Σ_a, typically a 5-μs burst of 14-MeV neutrons is used, and after the end of the neutron burst the die-away of thermal neutrons or gamma rays is observed for 1 000 to 3 000 μs. In the Soviet Union thermal neutron detectors have been used primarily [10], while much of the work in the United States of America has been with scintillation crystals to detect thermal neutron capture gamma rays.

In the USA, neutron die-away logging is commercially available from two service companies and is called "neutron lifetime log" (NLL) by Dresser-Atlas Corp. [11] and "thermal decay time log" (TDT) by Schlumberger Company [12]. The log gives a response similar to the open-hole electrical resistivity measurement, with low readings opposite salt-water-saturated zones (high Σ_a) and high readings opposite oil saturation (low Σ_a). The neutron die-away log is used to find oil-water contacts and to determine quantitatively the fluid content when porosity is available from other data, such as a CNL neutron log. Water saturations computed from the log response are generally good to about ± 15% [13]. Slight variations in water salinity, shaliness, hydrocarbon or rock matrix neutron absorption properties can result in significant saturation uncertainties. A correction for shaliness can often be made based on the relative natural gamma-ray log readings in the zones of interest. Pulsed neutron logging, because it can be performed in cased wells, is supplying previously unobtainable information for evaluating reservoirs [14]. When spaced in time, these logs make it possible to deduce the nature of water encroachment and hydrocarbon depletion at any stage in the producing life of a well.

Recent developments in neutron die-away logging are the use of two gamma-ray detectors at different spacings from the neutron source and variable positioning of the two gates used to determine the slope of the die-away curve. The TDT tools utilize a system of moveable and expandable detection gates, variable pulsing frequency and pulse width [12]. The gates and pulsing regime are automatically adjusted as the log is being run to optimize the thermal decay time measurement to yield the most precise determination of τ for a given neutron source strength. An additional gate, delayed until most of the decay has taken place, permits correction for background. This system is referred to as the complete scale factor method and has been described in detail. The latest version of the tool, called TDT-K, uses two detectors at different spacings [15] to measure the die-away rates, which are dependent on spacing, porosity and Σ_a. Since the spacing dependence is known, the two die-away rates can be used to deduce both porosity and Σ_a.

For quantitative interpretation of neutron die-away logs the absorption cross-section of the rock matrix must be known. One approach is to obtain samples of the rocks of interest and measure their Σ_a values in the laboratory. A pulsed neutron source technique for making such measurements has been developed, although the sample size required (15 - 20 kg) is larger than the core and cutting samples usually available [16]. The method requires only one thermal neutron lifetime determination for a mixture of crushed rock saturated with pure water. The technique has been used to measure crushed rock samples from large models used for neutron die-away research. Results were obtained with the method for igneous and sedimentary rocks which outcrop in various parts of the United States of America. Most of the measured cross-sections exceed the cross-sections calculated for a given rock on the basis of its primary minerals. It was concluded that most of the rocks probably contain strongly absorbing trace elements such as boron or gadolinium.

Another approach to quantitative interpretation of neutron die-away logs has recently been proposed to overcome the inaccuracy of data on Σ_a of the rock matrix [17, 18]. The procedure is to run the die-away log where the formation water salinity is relatively high and is known. Then fresh water is injected into the formation and the log is run again. With two sets of log data, both Σ_a of the rock and hydrocarbon saturation can be obtained. The objection to this philosophy is that it is applicable where only residual hydrocarbon is left in the formation. Otherwise, one reduces the hydrocarbon saturation by flushing it during fresh-water injection. The major problem is the error introduced by the fact that it is very difficult to achieve 100% displacement of formation water by the injected fresh water. In a similar technique slightly different steps are taken [18]. After running the pulsed neutron log, oil is removed by chemical flooding, then the "cleaned" formation is resaturated with its original water and another log is run. The problem of the flushing action remains with this second method.

B. Spectral gamma-ray logs

During the past several years improvements have been made in scintillation detectors, photomultiplier tubes, and solid-state electronics to the point that scintillation gamma-ray spectra obtained with borehole equipment rival the quality of spectra taken under laboratory conditions. This is primarily due to the development of integral combinations of NaI detectors and PM tubes that are stable at high temperatures, and to pulse amplifiers and electronic drivers for logging cables that exhibit good long-term stability. These improvements have made it possible to perform several types of spectral measurements with logging tools that have significant promise in solving certain geological, geophysical, and reservoir engineering problems.

Natural gamma-ray spectral logging

In principle, the simplest in-situ spectral measurement to make is that of radiation from gamma-ray emitting isotopes occurring naturally in earth formations. In the USA, natural gamma-ray spectral logging service is offered by two companies and has proved useful for correlation and mineral exploration and has potential applications for lithology identification. The log records simultaneously the total gamma-ray counting rate and the separate contributions from potassium and the decay products of uranium and thorium. The logging tool and uphole panel measure selectively gamma rays of a particular energy or a particular range of energies indicative of the isotope to be detected. Spectrum stripping techniques are used with a miniature computer in the logging truck to give the relative abundances of thorium, uranium and potassium. Calibration of the log is by means of three cylinders each containing a known amount of one of the elements potassium, uranium or thorium.

A typical log is shown in Fig. 3. On the left side are a self-potential electrical log and a conventional total count gamma-ray log made in 1947 when the well was drilled. In the centre is a new (1974) gamma-ray log and on the right side are the three spectral log traces of K, U and Th. The difference in the two total count gamma-ray logs is attributed to deposits of radioactive materials on the casing wall made over the years as the well

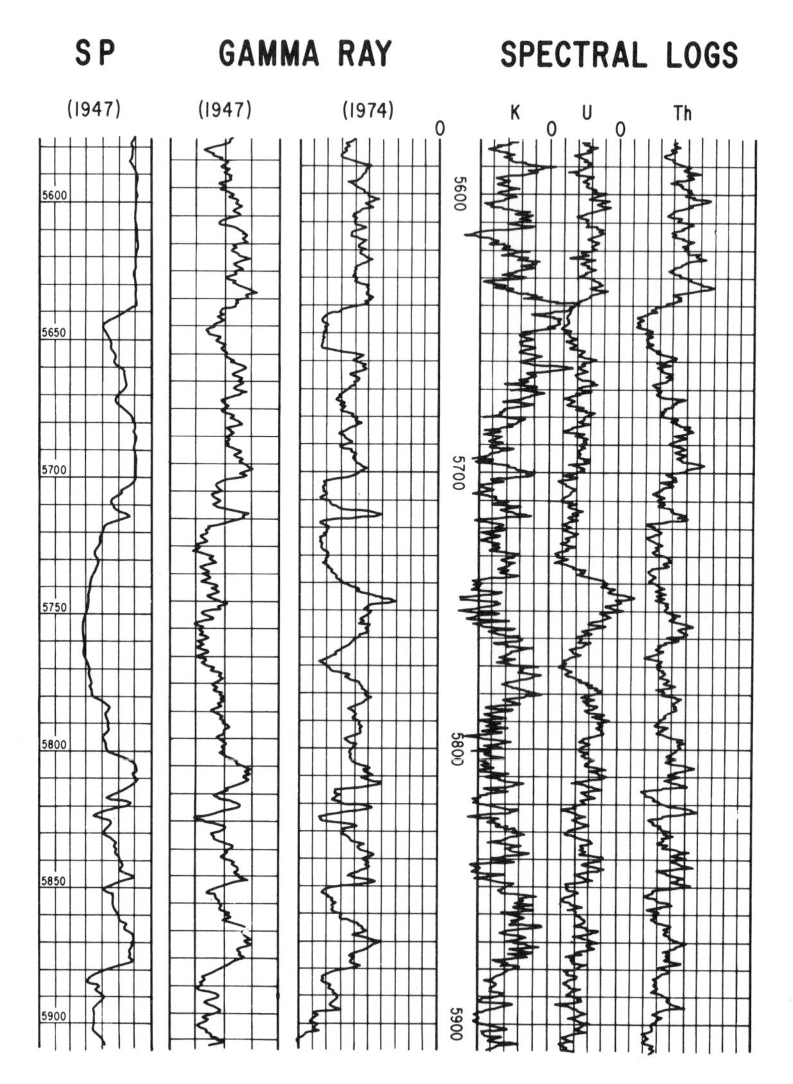

FIG. 3. Natural radioactivity logs.

was produced. The spectral traces show the nature of this deposit; for example, at 5745 ft the higher counting rate is due to uranium. In newly drilled wells the logs could readily be used to compute thorium-to-uranium ratios, which have been suggested as indicators of sedimentary processes since thorium is precipitated more readily than uranium in sea water.

Lock and Hoyer [19] have described the successful use of the spectral log to solve difficult correlation problems in the Santa Barbara Channel off the coast of California. The abundance of thorium in volcanic ash in a number of formations provided correlative time markers. They also described the use of the spectral log to identify potassium in exploration for potash deposits in the extensive evaporites in southeastern New Mexico.

Since the evaporite sequence contained very few radioactive clays, the thorium and uranium contents were low. Hence a quantitative measure of potassium content could be made from the counting rate in the 1.46-MeV gamma-ray peak. Lock and Hoyer also discuss other possible geological information that might be obtained from spectral measurements, including correlating facies changes with thorium-uranium ratios to reflect salinity of a basin, oxidation-reduction regime, and periods of transgression and regression. In uranium exploration the spectral logs have not been particularly useful because of radioactive disequilibrium of ores.

Carbon oxygen logging

The possibility of logging directly for hydrocarbons by the detection of carbon has been known for many years [20]. Since carbon emits a distinctive 4.43-MeV gamma ray from inelastic neutron scattering when it is bombarded with neutrons of sufficiently high energy, inelastic gamma ray spectroscopy would appear to be a promising approach. Early studies with isotopic neutron sources indicated that the carbon gamma ray suffered too much interference from capture and activation gamma rays produced in other elements to make this a viable method in practice. However, with the advent of pulsed neutron generators capable of being packed in a borehole configuration, the inelastic method of carbon detection was revived. This is due to the fact that capture and activation gamma rays can be discriminated against on the basis of time. Inelastic gamma rays occur coincidentally with the neutron burst during which time they are much more intense than capture or activation gamma rays. Therefore, a spectral measurement carried out during a time interval such as Gate 1 in Fig.2 will be representative primarily of inelastic gamma rays.

Within the last few years operational logging tools have been developed to carry out time-gated spectral measurements [21-23]. It has been found most satisfactory to determine the ratio of carbon to oxygen. If carbonates are present, a correction must be determined and applied to the measured C/O ratio to remove the contribution from carbon in the rock. This is done by measuring the Ca/Si ratio, or its inverse. By establishing empirical relationships in the laboratory and in known field cases, it is possible to determine hydrocarbon saturations in both sandstone and carbonate formations. An independent measurement of porosity is also required for this determination.

Carbon/oxygen logging will be most useful in those cases where the formation water salinity is very low or very poorly known. Experience thus far with the log has shown that it is a reliable indicator of hydrocarbons in sandstone formations. It is independent of formation water salinity, and seems to have only a weak dependence on formation shaliness. The Ca/Si ratio used with the C/O ratio has proven to be useful for differentiating limestones and limey sands from sandstones that contain hydrocarbons. In some cases high porosity gas sands can be distinguished from tight limestones.

Capture gamma-ray logging

The spectrum of gamma rays produced when neutrons are captured in earth formations contains useful information. Many elements have large capture cross-sections and emit intense gamma rays characteristic of the capturing nucleus. If capture gamma-ray spectroscopy could be carried out with sufficient accuracy in the borehole, information not heretofore available could be obtained.

In the past there have been some logs that are based on capture gamma rays. The response of neutron-gamma logs with isotopic neutron sources is primarily due to neutron capture in hydrogen. Chlorine logs (or salinity logs) have usually employed a crude form of spectroscopy in an attempt to isolate and measure the response to chlorine in the formation water. Neither type of log mentioned was a satisfactory spectral measurement, mainly because of interference from inelastic and activation gamma rays. This interference can be largely overcome by the use of a pulsed neutron generator. As shown in Fig. 2, the gamma rays present in the decay immediately following a neutron burst consist mostly of those from neutron capture plus a smaller contribution of activation gamma rays. If a spectral measurement is carried out during a time interval such as Gate 2, the quality of the capture spectrum should be greatly enhanced over that with an isotopic source.

Activation logging

Techniques are being developed for using accelerators and ^{252}Cf as neutron sources for borehole activation analysis. A number of experimental measurements have been made in laboratory models and in boreholes. No commercial service of substantial significance has yet developed. However, these techniques appear to be quite promising for lithology determination in many problem areas.

A continuous activation log is made by moving the neutron source ahead of the detector spaced far enough away from the source to avoid measuring capture gamma rays. The delayed gamma rays from nuclei activated by neutron bombardment are generally measured with a scintillation crystal detector so that energy analysis can be performed. Among the many elements which can be activated in the borehole, aluminium and silicon have been recorded in continuous logging runs with a borehole accelerator [24] and with capsule neutron sources [25]. The aluminium and silicon contents are helpful in defining productive intervals in shaly sands. The results cannot be interpreted uniquely since neutrons can produce ^{28}Al in shales from thermal neutron capture by aluminium or fast neutron capture by silicon. ^{252}Cf neutrons have a lower average energy of 2.3-MeV so that fewer neutrons from this source exceed the silicon threshold, giving less silicon interference than Pu-Be or 14-MeV neutrons.

Cyclic activation analysis has been developed to utilize very efficiently a low-output, pulsed source of 14-MeV neutrons and cyclic counting of induced activities [26]. The technique is particularly applicable to short-lived radioactivities — those with half-lives of a few minutes to as short as a few milliseconds. This technique differs from conventional activation with repeated mechanical transfer of a sample between the irradiating source and the detector. In cyclic activation the sample is fixed and cycling is electronic rather than mechanical. Cumulative detector response to radiations from induced activities is achieved by a succession of bombard-wait-count cycles. Counting is done between neutron bursts after the die-away of capture gamma rays. In laboratory experiments with silicate rocks, oxygen (7.35 s) and silicon (2.3 min) activities were prominent. This technique shows considerable promise for continuous logging of these short-lived activities.

Solid-state detectors

Lithium-drifted germanium detectors can provide energy resolution 10- or even 100-fold better than obtainable with scintillation crystal detectors. Such high resolution (~ 2 MeV at 1 MeV) should make it possible to identify a number of elements by their gamma-ray spectra. A few measurements have been made in shallow holes using a Ge(Li) detector.

The problem of cooling a Ge(Li) detector to a sufficiently low temperature for a length of time comparable to a typical logging run has not been solved in an operational logging tool. At present the most promising possibility seems to be the use of an intrinsic Ge detector. This type of detector does not have to be cooled in storage; it would require cooling only during the time of a measurement.

C. Uranium logging

A paper by Czubek in 1972 [27] is directed to uranium detection based on the use of a pulsed source of neutrons. The method employs measurement of neutron time distributions in rock media following a burst of fast neutrons. The paper presents a time- and energy-dependent theory for the four principal time distributions occurring after production of fission neutrons by prompt and delayed reactions and concludes that the detection of uranium in rocks is possible using either the epithermal neutron time distribution from the prompt fission of ^{235}U by thermal neutrons or the delayed neutron distribution from fast neutron fission of ^{238}U or thermal neutron fission of ^{235}U. Experiments performed by Czubek at CEN-Saclay in laboratory models of uranium ore referred to in the paper show the feasibility of detection of fission neutrons and the validity of the theory. The author concluded that, for normal borehole logging measurements, a pulsed neutron source with an average output of about 10^9 n/s is needed.

A US patent [28] issued in 1972 on an application filed in 1970 discloses a technique for borehole assay for uranium by measurement of delayed fission neutrons produced by 14-MeV neutron bombardment of the formations. A logging tool containing a pulsed neutron source and a neutron detector are located in a borehole at the level of a formation of interest as shown in Fig.4. The source is operated cyclically to irradiate the formation with bursts of fast neutrons, and the resulting neutrons from the formations are detected. Measurements are made between the neutron bursts and after the source neutrons have decayed and are indicative of delayed neutrons emitted as a result of neutron fission of uranium. Measurements are also obtained in a non-ore-bearing formation to record the count of delayed neutrons emitted from oxygen when irradiated. These measurements are compared with those obtained in the ore-bearing formations of interest to correct for the effect of the oxygen background due to delayed neutrons from ^{17}O. Although thorium may also emit delayed neutrons the effect normally is very small. The delayed fission neutrons from uranium come primarily from six neutron groups with half-lives ranging from 0.23 to 55.7 s, the most abundant group having a half-life of about 2.5 s. In order to ensure

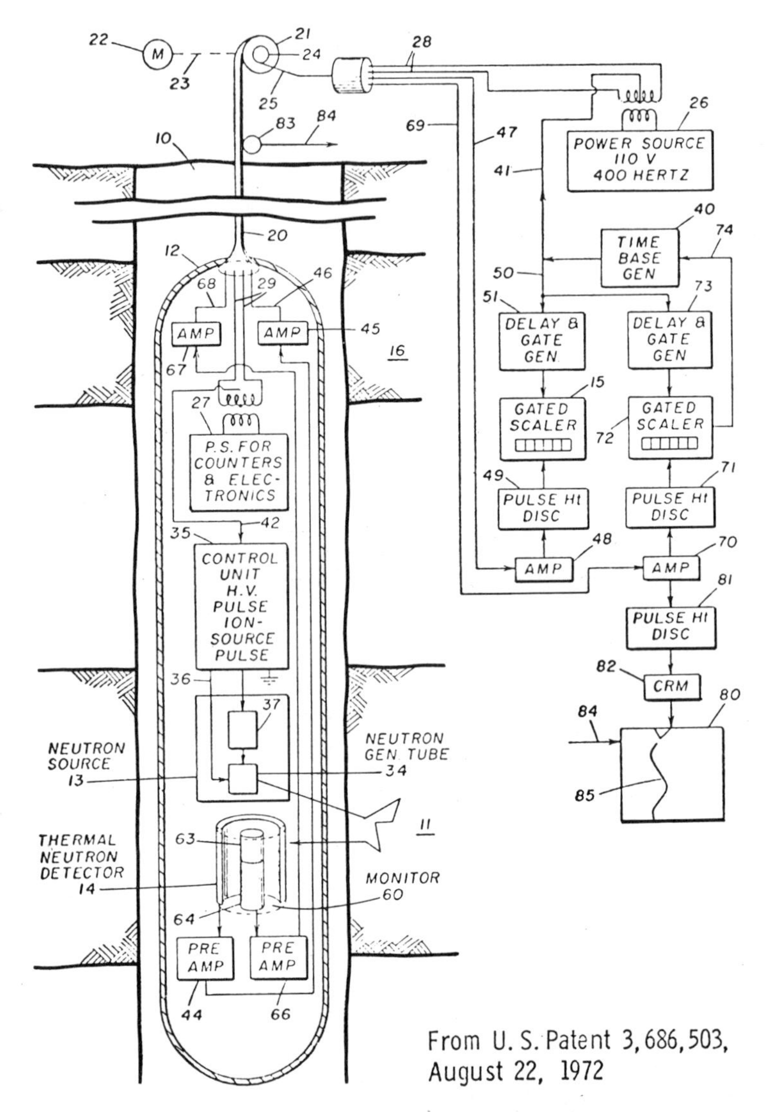

FIG. 4. Delayed fission neutron logging for uranium.

that no moderated source neutrons are counted, the scaler associated with the neutron detector is operated to begin counting at 5 milliseconds after each neutron burst and counts until the beginning of the next neutron burst. The source may be operated at 2 bursts a second, each burst having a duration of about 3 microseconds. In commercial ore zones an assay can be made by counting delayed neutrons for about a 5-min period. This technique has been successfully employed in field logging in the United States of America utilizing a neutron tube manufactured by Kaman Nuclear Sciences in a logging tool designed along the lines of the patent.

IV. DENSITY LOGGING

Formation density logging based on gamma-ray scattering and absorption became available commercially in 1954. An intermediate-energy gamma-ray source is used, usually ^{137}Cs. For the energy range 0.5 to 2.0 MeV, Compton scattering predominates in the scattering and absorption of gamma rays. By using a collimated source of gamma rays and a collimated scintillation crystal detector, narrow beam geometry can be approximated in a logging tool and hence exponential absorption applies. For about 10 years a relatively simple logging device was used in which the whole tool was pushed against the borehole wall by a bow spring. The source and a single scintillation crystal detector were located near the bottom of the tool. Lead shielding reduced the direct transmission of gamma rays either through the instrument or through the well bore fluid, but the collimation was not well defined. The response of the tool was seriously affected by mudcake and borehole irregularities.

A compensated formation density logging device (FDC) with two detectors at different distances from the source was introduced in 1964 [29]. The source and detectors are in a skid on a moveable arm which presses the skid firmly against the borehole wall. The detector nearer to the gamma-ray source is primarily sensitive to the density of material immediately adjacent to the face of the skid, whereas the far detector is more sensitive to the bulk of rock beyond the borehole wall. The signals from both detectors are combined to give automatically a density correction for mudcake and minor borehole irregularities and the corrected density from the reading from the detector at the longer spacing. Both the compensated density measurement and the amount of compensation are recorded on the log. The compensated density reads directly in grams per cubic centimetre. Primary calibration standards are laboratory fresh-water-filled limestone formations; secondary standards are large aluminium and sulphur blocks.

The principal use of the density log is to calculate formation porosity and hence fluid content. The density log is available with a formation porosity index scale for either sandstone or limestone lithology. The lithology must be known (from other logs or core data) for the most accurate porosity determinations. For a single unknown lithology, one can use a plot of density log response versus resistivity to obtain grain density to use in the well-known equation for computing porosity from density data. When the rock matrix is composed of two lithologies, one can plot density log versus neutron log responses to determine the proportion of the two lithologies and hence determine the correct porosity. Density data are also used by the geophysicists in seismic modelling studies. The combination of the gamma-gamma density log and borehole gravity metre measurements of bulk density are used for exploration purposes. The compensated density log is also used in combination with the compensated neutron log for locating gas, as has been described in Section II-C.

Density logs are also useful for locating coal, for determining its bed boundaries and for estimating ash content of coal [30]. The low densities of coals — from 1.4 to 1.8 g/cm^3 for anthracite, 1.2 to 1.5 g/cm^3 for bituminous, and 0.7 to 1.5 g/cm^3 for lignite — cause distinctive log deviations. The ash content of coal, which depends on the quantity of included mineral impurities, such as silica, alumina, and iron oxide, is an important factor in determining its commercial value. Because all mineral impurities of significance have appreciably higher densities than the carbon and hydrocarbons of coal, a density log provides an estimation of the variations in ash content for any given coal seam. The ash content is related very nearly linearly to specific gravity or density within a given coal seam [31]. The accuracy of the density logs run for coal is generally not as good as that of the logs run in petroleum exploration, because the holes drilled for coal are smaller in diameter and hence the logging tool is smaller in diameter with less sophisticated collimation of the source and a single detector.

In 1952, Bardsley and Algermissen showed the usefulness of acoustic and density logs for evaluating oil shale. They used commercial logging tools in conventional size holes to get good oil yield values compared with core assay. In 1966, Baldwin et al. [32] demonstrated that good oil yield values could be obtained in small-diameter (7.62 cm) holes using a small-diameter (4.45 cm) density tool. The logs were calibrated using density results reported in the literature for oil shale and borehole models of known density. Reasonable values of oil yield were obtained. Later the log results were compared with assay analysis of cores taken from the holes logged and good agreement was found. The use of small-diameter density logging tools in oil shale has been very limited because of the small commercial market. But the situation is changing as interest in oil shale increases with the sharp rise in the price of hydrocarbons.

REFERENCES

[1] WORKING GROUP ON NUCLEAR TECHNIQUES IN HYDROLOGY, Nuclear Well Logging in Hydrology, Technical Reports Series No. 126, IAEA, Vienna (1971).

[2] MILLS, W.R., CALDWELL, R.L., MORGAN, I.L., Low voltage ^{3}He-filled proportional counter for efficient detection of thermal and epithermal neutrons, Rev. Sci. Instrum. 33 8 (1962) 866.

[3] TITTMAN, J., SHERMAN, H., NAGEL, W.A., ALGER, R.P., The sidewall epithermal neutron porosity log, J. Pet. Technol. 18 10 (1966) 1351.

[4] ALLEN, L.S., TITTLE, C.W., MILLS, W.R., CALDWELL, R.L., Dual-spaced neutron logging for porosity, Geophysics 32 1 (1967) 60.

[5] ALGER, R.P., LOCKE, S., NAGEL, W.A., SHERMAN, H., The dual spacing neutron log - CNL, Soc. Pet. Eng. of AIME, New Orleans (1971).

[6] TRUMAN, R.B., ALGER, R.P., CONNELL, J.G., SMITH, R.L., Progress report on interpretation of the dual-spacing neutron log (CNL) in the U.S., Soc. Pet. Eng. of AIME, New Orleans (1971).

[7] ALLEN, L.S., MILLS, W.R., DESAI, K.P., CALDWELL, R.L., Some features of dual-spaced neutron porosity logging, Soc. Prof. Well Log Analysts, Tulsa (1972).

[8] CALDWELL, R.L., MILLS, W.R., ALLEN, L.S., BELL, P.R., HEATH, R.L., Combination neutron experiment for remote analysis. Science 152 3721 (1966) 457.

[9] MILLS, W.R., GIVENS, W.W., CALDWELL, R.L., "Water analysis by a combination neutron experiment", Geol. Problems in Lunar and Planetary Res., Sci. and Tech. (published by Am. Astronautical Soc.) 25 (1971) 185.

[10] CALDWELL, R.L., Well logging in the U.S.S.R., The Log Analyst 9 2 (1968) 16.

[11] YOUMANS, A.H., HOPKINSON, E.C., BERGAN, R.A., OSHRY, H.I., Neutron lifetime, A new nuclear log, Soc. Pet. Eng. of AIME, New Orleans (1963).

[12] WAHL, J.S., NELLIGAN, W.B., FRENTROP, A.H., JOHNSTONE, C.W., SCHWARTZ, R.J., The thermal neutron decay time log, Soc. Pet. Eng. of AIME, Houston (1968).

[13] STIEBER, S.J., Pulsed neutron capture log evaluation — Louisiana Gulf Coast, Soc. Pet. Eng. of AIME.

[14] FONS, L., Some pulsed neutron logging contributions to improved formation evaluation, J. Pet. Technol. 22 (1970) 424; THREADGOLD, P., Interpretation of thermal neutron die away logs — some useful relationships. Soc. Prof. Well Log Analysts, Dallas (1971).

[15] DEWAN, J.T., JOHNSTONE, C.W., JACOBSON, L.A. WALL, W.B., ALGER, R.P., Thermal neutron decay time logging using dual detection, Soc. Prof. Well Log Analysts, Lafayette (1973).
[16] ALLEN, L.S., MILLS, W.R., Measurements of the thermal neutron absorption cross section of rock samples by a pulsed source method, Soc. Prof. Well Log Analysts, McAllen (1974).
[17] MURPHY, R.P., OWENS, W.W., The use of special coring and logging procedures for defining reservoir residual oil saturations, J. Pet. Technol. 25 (1973) 841.
[18] RICHARDSON, J.E., WYMAN, R.E., JORDEN, J.R., MITCHELL, F.R., Method for determining residual oil with pulsed neutron capture logs, J. Pet. Technol. 25 (1973) 593.
[19] LOCK, G.A., HOYER, W.A., Natural gamma-ray spectral logging, The Log Analyst 12 5 (1971) 3.
[20] CALDWELL, R.L., SIPPEL, R.F., New developments in radioactive well logging research, Bull. Am. Assoc. Pet. Geol. 42 1 (1958) 159.
[21] LOCK, G.A., HOYER, W.A., Carbon-oxygen (C/O) log: use and interpretation, Soc. Pet. Eng. of AIME, Las Vegas (1973).
[22] CULVER, R.B., HOPKINSON, E.C., YOUMANS, A.H., Carbon oxygen (C/O) logging instrumenfation, Soc. Pet. Eng. of AIME, Las Vegas (1973).
[23] SMITH, H.D., Jr., SCHULTZ, W.E., Field experience in determining oil saturations from continuous C/O and Ca/Si logs independent of salinity and shaliness, Soc. Prof. Well Log Analysts, McAllen (1974).
[24] WICHMANN, P.A., WEBB, R.W., Neutron activation logging for Si to Al ratios, J. Pet. Technol. 22 2 (1970) 201.
[25] PAAP, H.J., SCOTT, H.D., The use of ^{252}Cf as a neutron source for well logging, Symp. on Applications of Radioisotopes, Am. Inst. of Chem. Eng., Chicago (1970).
[26] GIVENS, W.W., CALDWELL, R.L., MILLS, W.R., Cyclic activation logging, The Log Analyst 9 3 (1968) 18.
[27] CZUBEK, J.A., Pulsed neutron method for uranium well logging, Geophysics 37 1 (1972) 160.
[28] GIVENS, W.W., CALDWELL, R.R., MILLS, W.R., In situ assaying for uranium rock formation, US Patent ≠ 3,686,503, Aug. 22, 1972.
[29] WAHL, J.S., TITTMAN, J., JOHNSTONE, C.W., ALGER, R.P., The dual spacing formation density log, J. Pet. Technol. 16 (1964) 1411.
[30] CALDWELL, R.L., Nuclear logging methods, Isot. Radiat. Technol. 6 3 (1969) 257.
[31] TIXIER, M.P., ALGER, R.P., Log evaluation of non-metallic mineral deposits, Soc. Prof. Well Log Analysts, Denver (1967).
[32] BALDWIN, W.F., CALDWELL, R.L., GLENN, E.E., HICKMAN, J.B., NORTON, L.J., Slim hole logging in Colorado oil shale, Soc. Prof. Well Log Analysts, Tulsa (1966).

BRIEF REVIEW OF DEVELOPMENTS IN NUCLEAR GEOPHYSICS IN SWEDEN

R. CHRISTELL, K. LJUNGGREN
Isotope Techniques Laboratory,
Stockholm

O. LANDSTRÖM
AB Atomenergi,
Studsvik, Sweden

Abstract

BRIEF REVIEW OF DEVELOPMENTS IN NUCLEAR GEOPHYSICS IN SWEDEN.

The geology of Swedish ore deposits is described briefly. The most important ores in Sweden are iron ores and sulphide ores, the latter containing mainly copper, zinc and lead. There is also a large low-grade uranium deposit in alum shale. The variety of ore deposits and ore elements calls for flexible, multipurpose logging techniques. A research and development programme for developing methods and instrumentation to meet this need was undertaken jointly by AB Atomenergi and ITL. The measurements considered were: (A) continuous logging of boreholes for (a) natural radioactivity, (b) rock density (by γ-γ techniques), (c) heavy mineral zones (by selective γ-γ techniques), and (d) elemental concentrations (by X-ray fluorescence, neutron capture, and activation analysis for short-lived nuclides); as well as (B) stationary measurements at selected positions in boreholes for elemental concentrations (by natural gamma spectrometry, X-ray fluorescence, activation, and γ-γ methods). A brief description is given of the methods and equipment used. Copper logging by means of neutron activation for ^{64}Cu and lead logging by means of X-ray fluorescence, in the presence of barium, are reported in some detail.

1. INTRODUCTION

1.1. Geology of Swedish ore deposits

The Swedish bedrock contains significant concentrations of many metals and is especially rich in iron ores. Sulphide deposits, mainly copper, zinc and lead, also constitute important ores.

The most important iron ores are found in two ore regions, one in northern Sweden (the Kiruna and Gällivare district) and the other in central Sweden (Bergslagen). Important mines in the former region are Kiirunavaara (the largest ore body in Sweden), Malmberget, Leveäniemi (Svappavaara) and Tuollavaara. In central Sweden there are several iron-ore mines of varying size, the most important ones being Grängesberg and Stråssa. The iron ores, which occur in Precambrian supracrustal rocks in both regions, are composed of iron oxides, mainly as magnetite and in some mines also as hematite.

The most important sulphide ores occur in four ore provinces: northern Sweden, the Skellefte district, the Caledonian mountain chain and central Sweden (Bergslagen).

In northern Sweden (Aitik in the Gällivare district), mining of a low-grade copper ore was started in recent years. This ore is of a porphyry copper type.

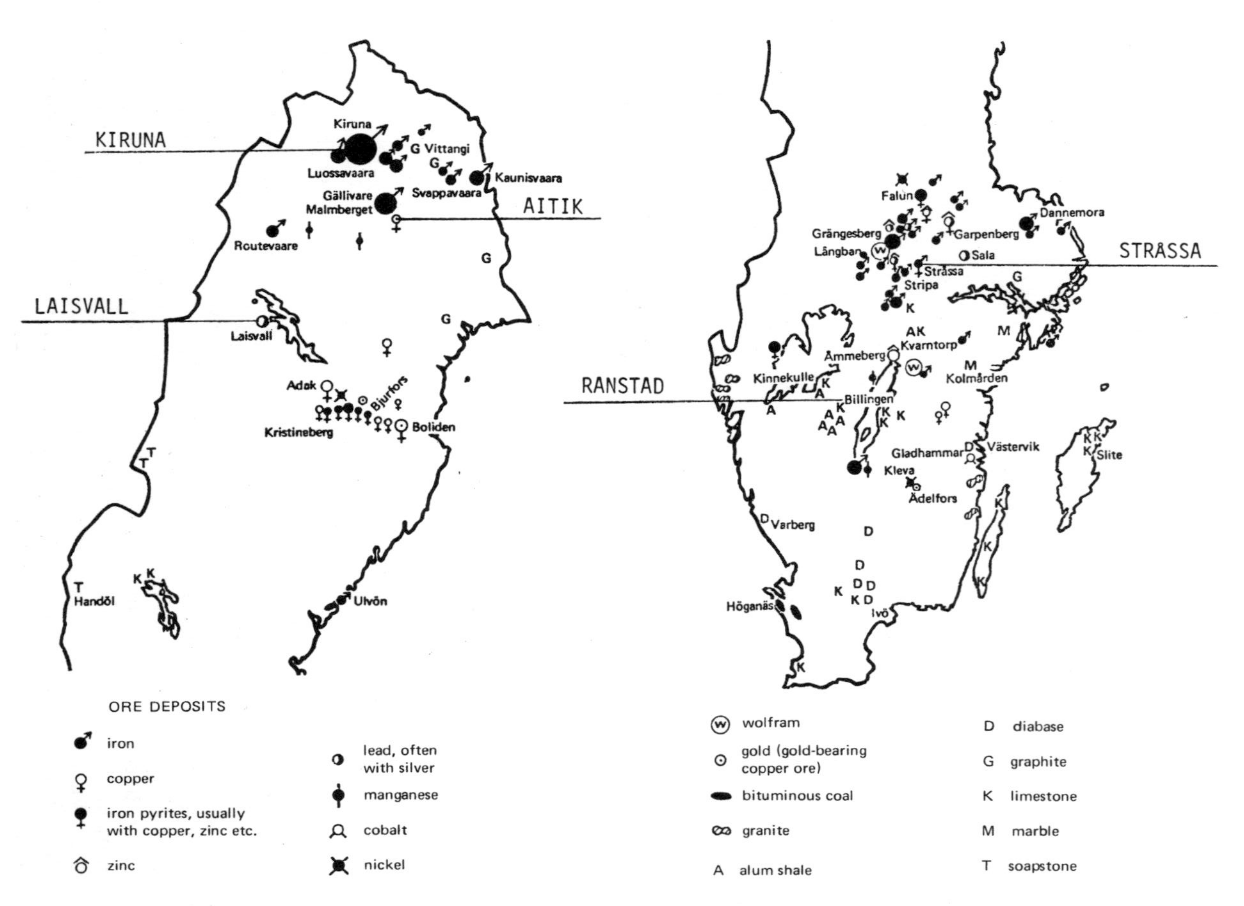

FIG.1. Locations of ores and mines in Sweden. The localities where nuclear techniques have been tested in the field are indicated.

The Skellefte district is very rich in sulphide ores, the main ore metals being copper, lead and zinc. Minor ore elements are gold and silver. The most important ores mined at present are at Kristineberg, Renström and Adak.

In the Caledonian mountain chain a number of lead-zinc deposits occur in Eocambrian and Cambrian sandstones and quartzites along the eastern border. The most important ore deposit is Laisvall in northern Sweden which is mined for lead. In the inner part of the Caledonides (close to the Norwegian border) sulphide deposits, which also carry copper, are found. Of these, the copper ore at Stekenjokk is worth mentioning.

In central Sweden (Bergslagen) there are several sulphide deposits. The most important of these being mined at present are Åmmeberg and Garpenberg, which are mined primarily for zinc and lead. The latter deposit also contains small amounts of copper. Tungsten is at present extracted only in the Yxsjöberg mine.

The only uranium ore mined in Sweden is located at Ranstad in southern Sweden. The ore occurs as alum shale of Cambrian age with an ore grade of 200-300 g uranium per ton. Figure 1 shows the locations of the most important ore deposits in Sweden.

The economic importance of the abundant ore deposits in Sweden has called for the development and adoption of various methods for measuring the concentration of essential elements in rock for prospecting and mining purposes. Among these methods, nuclear methods have gained increasing importance both for the determination of natural radioactivity, especially in the search for uranium, and for in-situ analysis of other elements, particularly iron, copper, lead and zinc. Even if the elements mentioned are the most important ones in the Swedish mining industry, there are several others for which in-situ analysis may be of great value, e. g. tungsten, silver tin, nickel and chromium. A research and development programme in nuclear geophysics, initiated in 1967, has been jointly executed by AB Atomenergi and ITL.

1.2. Required techniques and objectives

The large number of different kinds of ore deposits (and ore elements) in the Swedish bedrock gives rise to the need for flexible, multipurpose logging techniques. To assist mining companies and geological institutes in their exploration and exploitation work, logging equipment needs to be designed in such a way that it can be used for several types of in-situ borehole measurements with only minor adjustments between runs. The following measurements were considered:

(A) Continuous logging of boreholes for:

- (a) natural radioactivity
- (b) rock density (by γ-γ techniques)
- (c) heavy mineral zones (by selective γ-γ techniques)
- (d) elemental concentrations (by X-ray fluorescence, neutron capture, and activation analysis of short-lived nuclides);

(B) Stationary measurements at selected positions in boreholes for:

- (a) elemental concentrations (by natural gamma spectrometry, X-ray fluorescence, activation, and γ-γ methods).

TABLE I. SURVEY OF INVESTIGATIONS IN SWEDISH MINES

Methods	Spectrometric measurements of natural gamma radiations	Neutron activation analysis	Measurement of back-scattered gamma radiations		Measurement of gamma radiations from neutron capture	Measurements using radioisotope X-ray techniques
			"Soft" source	"Hard" source		
Elements	K, Th, U	Ag, Al, Ba, Cu, Mn, Na	Ba, Fe, Pb	Fe, Pb Rock density	Fe	Pb, Ba
Mines	Aitik (Boliden AB) Kiruna (LKAB) Laisvall (Boliden AB) Ranstad (AB Atomenergi) Stråssa (Gränges AB)	Aitik Stråssa	Laisvall Kiruna	Laisvall Kiruna Stråssa	Leveäniemi (LKAB) Stråssa	Laisvall
Radioisotope sources		^{210}Po-Be (10 Ci)	^{75}Se (0.3 mCi)	^{137}Cs (10 mCi) ^{60}Co (2 mCi)	^{210}Po-Be (10 Ci)	^{51}Cr (0.1 mCi) ^{75}Se (0.3 mCi) ^{137}Cs (10 mCi) [^{60}Co (2 mCi)]

The metals of primary interest in the project were copper, iron and lead because of their abundance and economic significance, but other elements of interest in the Swedish mining industry were considered as well.

1.3. Survey of methods used

Activation analysis by means of an isotopic neutron source was used for determining copper. Neutron-capture gamma-ray logging (n-γ) and gamma back-scattering (γ-γ) were used for determining iron. This latter method was also used for determining lead and rock density. X-ray fluorescence techniques have also been successfully used for determining lead. Uranium, thorium and potassium were determined by spectrometry of natural gamma radiation. The measuring methods used and the test localities are given in Table I.

The results of the attempts to determine copper in the Aitik mine indicated that activation analysis would permit in-situ borehole assays to be made. Because of the low copper content of the ore and the high radon content of the borehole water, which interferes with the copper activity, a stronger radiation source (10^8 n/s) would be required than that available for the exploratory measurements [1].

Determination of iron content and rock density by means of the γ-γ method has been tested on ores in the Kiruna field and in the Stråssa mine. The results have been compared with those obtained by conventional analysis of drill-cores. The method offers a simple means for determining ore boundaries.

The n-γ method, which makes use of the characteristic gamma radiation emitted promptly upon capture of thermal neutrons, has also been tested for iron determination in the Stråssa and Leveäniemi mines. This method has the advantage of being specific for the element sought which, in this case, is iron, but it only seems to be practicable at iron concentrations below 30%.

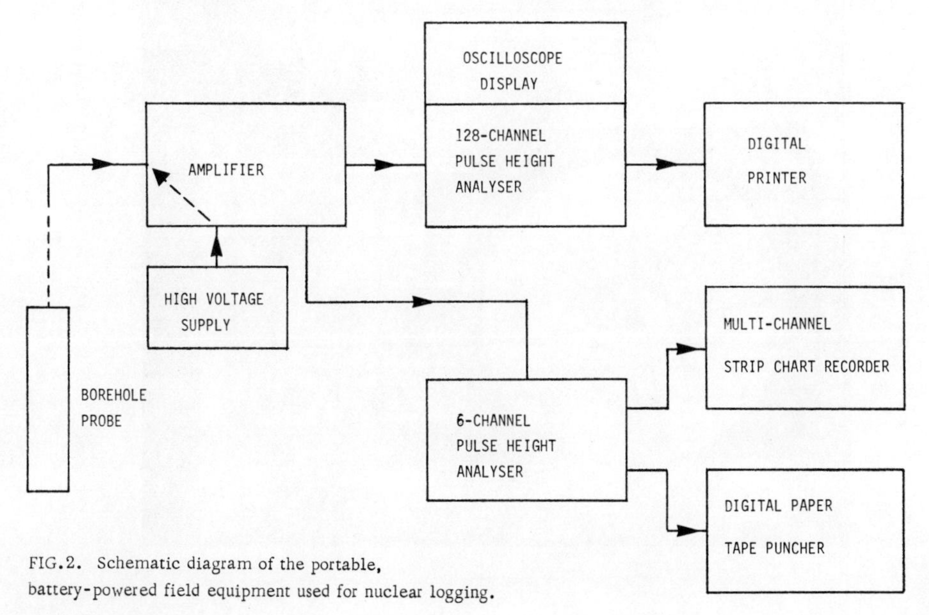

FIG.2. Schematic diagram of the portable, battery-powered field equipment used for nuclear logging.

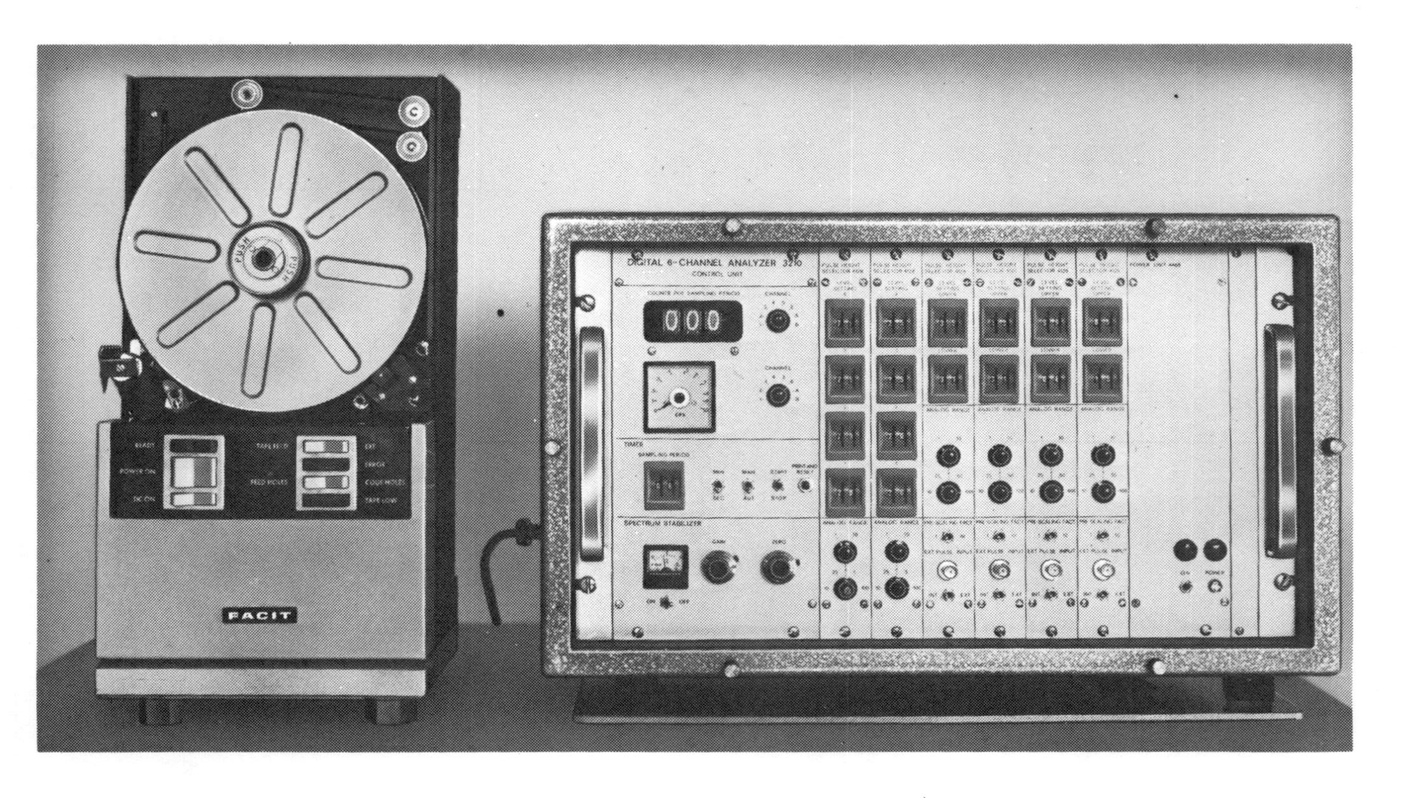

FIG.3. Six-channel spectrometer designed for multipurpose logging.

In Laisvall, gamma back-scattering techniques were used to determine lead and to localize heavy mineral zones. The results were promising, indicating that quantitative determinations of lead would most likely be possible provided that no other heavy minerals were present. Unfortunately, from this point of view, the ore frequently contained significant amounts of barium which rendered reliable lead determination impossible by γ-γ techniques.

Instead, the development work was concentrated on measurement of the characteristic X-rays of lead and barium. With this technique, both elements could be selectively determined with satisfactory accuracy. Interference from other heavy elements and matrix effects could be eliminated by means of a simple correction procedure.

In general, X-ray fluorescence techniques are capable of yielding concentration values of satisfactory accuracy. They can also be used for other heavy elements such as tungsten. For lighter elements, such as zinc, copper and iron, the low energy of the emitted X-rays limits the usefulness of the method.

Accurate quantitative determination of uranium at Ranstad has been made by means of spectrometry of natural gamma radiations. Logging of exploration boreholes at Laisvall and Kiruna and in the Aitik and Stråssa mines demonstrated that gamma spectrometry permits assay of uranium, thorium and potassium to be made even at the concentrations which occur in ordinary rock.

Some of the methods described, especially X-ray fluorescence and activation analysis, are specific for each element and may therefore be of great value in identifying elements and minerals in borehole zones where, due to core losses, no information on the composition of the rock or the ore is available by other means.

1.4. Equipment used

For the field investigations versatile, battery-operated, mobile logging equipment was required. Commercially available units were combined with units which were developed and designed to specifications such that the complete system would meet the actual requirements of performance. The equipment is schematically shown in Fig. 2. Two sizes of logging probes are used, one 43 mm in diameter (for 46-mm-dia. boreholes) and the other 75 mm in diameter (for $\geq$ 80-mm-dia. boreholes). Each probe contains a radiation source, which is exchangeable, radiation shields (absorbers between source and detector crystal), a sodium iodide scintillation detector and a pre-amplifier. The probes are connected by up to 1 000 m of self-supporting cable to the "ground" instrumentation, which consists of a main amplifier, a high-voltage supply, a gamma radiation spectrometer and recording units.

As a rule, only a few energy channels are needed for the actual measurements. A versatile six-channel spectrometer has been specially designed for this purpose (Fig. 3). It can be used in combination with various types of logging probes and for air-borne or car-borne surveys. However, it has also been found very practical to have access to a multi-channel spectrometer in the introductory stage of all logging investigations in order to get an overall display of the complete energy spectrum available for orientation.

A logging probe with a Ge(Li) detector, developed by AB Atomenergi, has since been added to the instrumentation system [2].

2. INVESTIGATIONS AND RESULTS

2.1. Natural radioactivity logging

Spectrometric measurements of natural gamma radiation allow the gamma lines characteristic of uranium, thorium and potassium to be identified. This fact can be utilized for determining these elements quantitatively. The determination is usually based on measurements in energy intervals at 1.46 MeV (potassium), 1.76 MeV (uranium) and 2.62 MeV (thorium). The count in the thorium channel is a direct measure of thorium, whereas the uranium channel has to be corrected for the contribution from thorium and the potassium channel for the contributions from both thorium and uranium.

The use of a Ge(Li) probe is likely to increase the accuracy and sensitivity of the measurement [2] because of the high resolution of the detector: the total absorption peak area can be easily determined (with the six-channel analyser), thus eliminating the subtracting methods necessary when a sodium iodide detector is used.

Spectrometric borehole measurements of natural radiation were always made in connection with in-situ activation analysis to obtain a knowledge of the natural background. Spectrometric measurements have also been applied to the determination of uranium, thorium and potassium in rocks. The data are easily converted to values of radioactive heat generation in the rock which are necessary for adequate interpretation of different geothermal investigations [3].

The concentrations of the radioactive elements are usually a characteristic property of different rocks and have been used for lithological identification in coreless holes. This is especially true of sedimentary rock sequences where the total counts are usually sufficient for such an identification.

Quantitative in-situ determination of uranium in the Ranstad alum shale can be made with a fairly simple gamma spectrometric method since radioactive equilibrium is established and the radon leakage is very low. Figure 4 shows the correlation between the intensity of the high-energy gamma rays and core analyses (the potassium gamma radiation is eliminated by discrimination). Chemical analyses have shown that the thorium content is low and relatively constant in the ore zone; it has been corrected for in the diagram.

The gamma spectrometric technique is, at present, used routinely only for aerial prospecting by the Swedish Geological Survey (SGU). Some prospecting is also carried out by private companies.

2.2. Logging by means of γ-γ techniques

Back-scattered gamma radiation can be utilized for density measurement and assay of heavy minerals. The two phenomena which control the interaction of gamma radiation with matter are photo-electric absorption and Compton scattering. The probability for photo-absorption increases strongly with increasing absorber atomic number. Low radiation energy also favours photo-absorption. By using a radiation source which emits low-energy gamma radiation one can achieve a strong correlation between

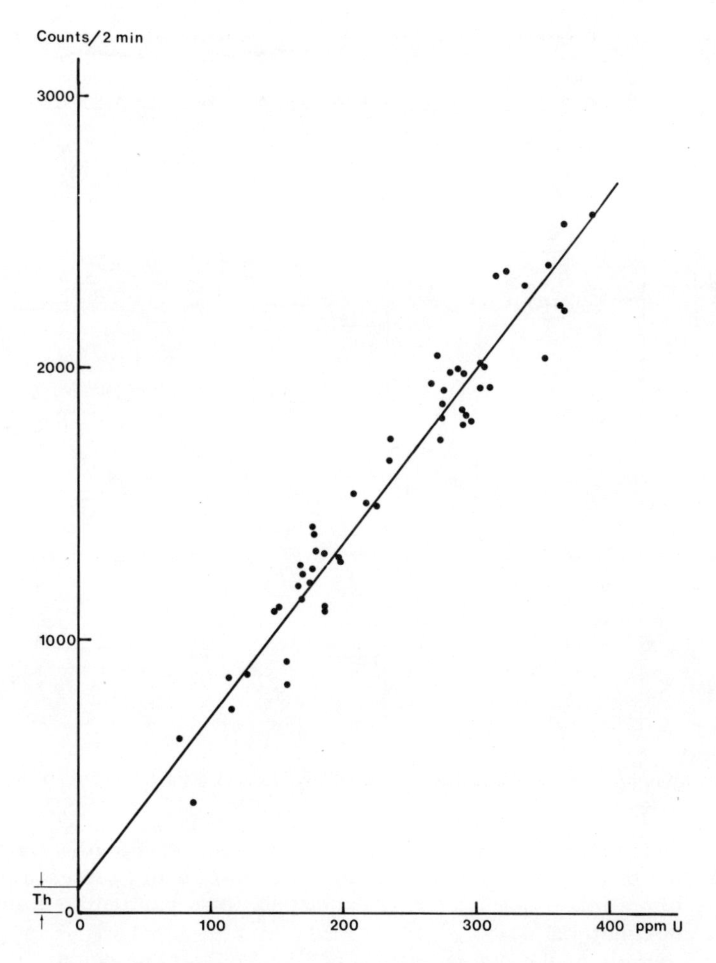

FIG.4. Calibration line for uranium in boreholes (Ranstad, natural gamma radiation logging).

the intensity of the back-scattered radiation and the content of elements of high atomic number in the rock. This method is sometimes referred to as selective γ-γ logging.

Compton scattering dominates when gamma radiation of medium energy interacts with elements of low to medium atomic number. For a given radiation energy, the Compton scattering efficiency is proportional to the electron density of the absorber. The electron density in turn depends on the density and on the ratio between the atomic number and the atomic weight. Since this ratio is nearly constant for different types of rock, the back-scattered radiation will be approximately proportional to the rock density (provided that the Compton-type interaction dominates). In some rocks the density is in turn determined by the concentration of one specific mineral which can then be assessed by means of density logging.

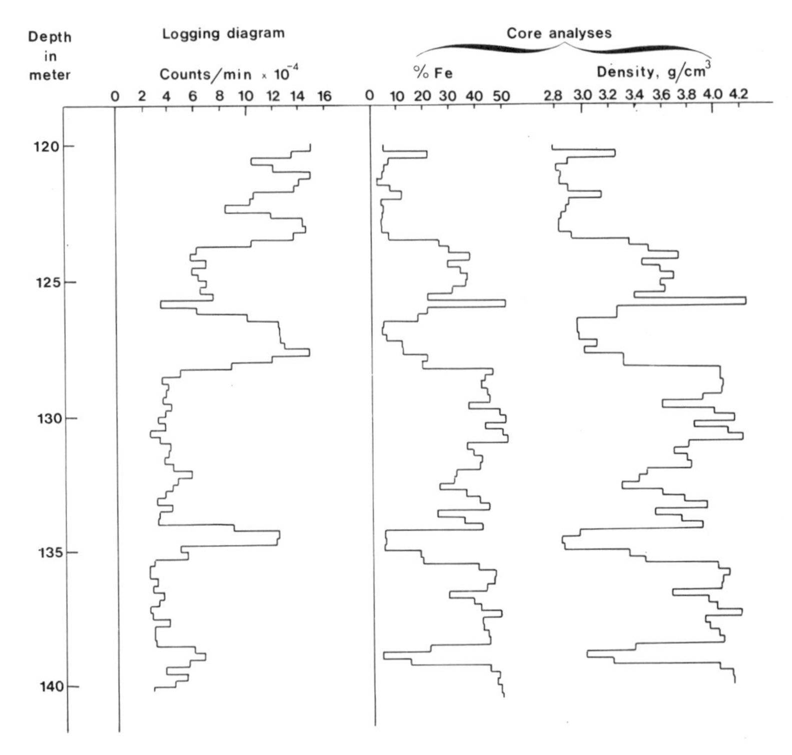

FIG. 5. Logging diagram and core analysis from an iron ore (Nukutus, Kiruna, γ-γ logging with ^{137}Cs).

By proper choice of radiation source and source/detector distance and by selection of back-scattered radiation in certain energy intervals it is possible to optimize the measuring conditions for a particular element in a given rock matrix.

An example of the determination of iron content and density in an iron ore is given in Fig. 5. The logging was made by means of a ^{137}Cs source in the Nukutus ore at Kiruna in core-drilled holes. The results were compared with those of the core analyses for corresponding depth intervals. The results are satisfactory with regard to resolution although very narrow strata of attle rock are hardly recognizable. Better results were obtained by using a low-energy radiation source (^{75}Se).

Another example are the loggings carried out in the Stråssa mine to investigate the possibility of determining iron content by means of a probe containing a ^{60}Co source. The back-scattered radiation was registered in two separate energy intervals, one covering 100-125 keV and the other covering energies above 125 keV. Comparisons between back-scattered radiation intensity in each energy interval and the results from analyses are presented in Figs 6 and 7.

The iron content figures in Fig. 6 were determined by averaging over rather long core segments, whereas those in Fig. 7 were determined by averaging over three 100-mm intervals (the position of the probe in the borehole relative to the core is shown).

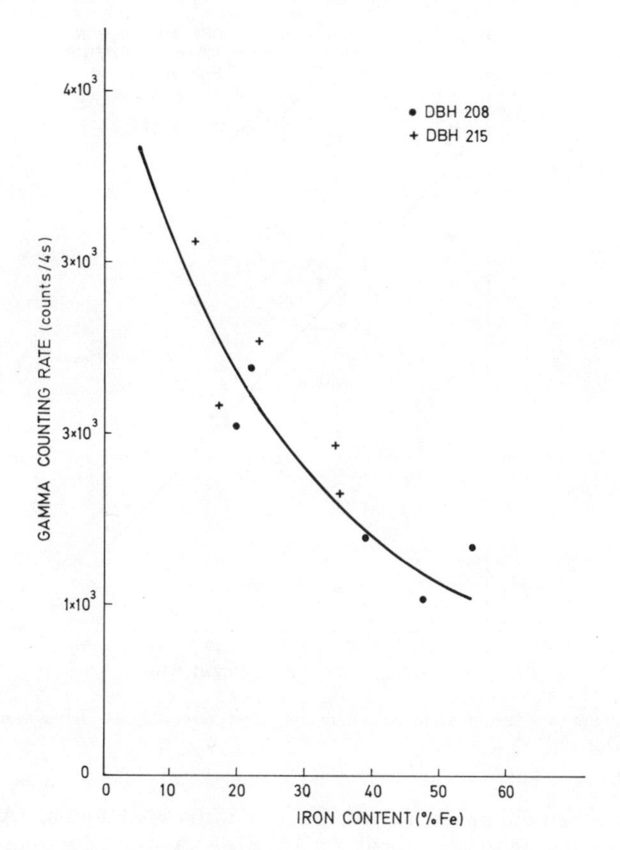

FIG.6. Intensity of back-scattered gamma radiation in the energy range 100-125 keV plotted against iron content in cores (γ-γ logging with ^{60}Co at Stråssa).

The results of the loggings scatter around a calibration line based on the loggings combined with the core analyses. The results indicate that no improvement will be obtained by measurement in other energy intervals or by modified averaging over the core segments. One obvious reason for the differing values is that the radiation penetrates rather deeply into the surrounding rock and therefore will yield values which are representative of a much larger volume than corresponds to the core which is only 36 mm in diameter. A "true" comparison can never be made. The logging therefore gives results that represent larger rock volumes than those represented by the cores and should best be correlated with ore estimates over larger volumes and longer time periods than were obtainable in the reported project.

The selective γ-γ method was found useful for the identification of heavy mineral zones.

2.3. Neutron activation logging

Field experiments were performed in the Aitik open pit copper mine to investigate the possibility of in-situ borehole determination of copper

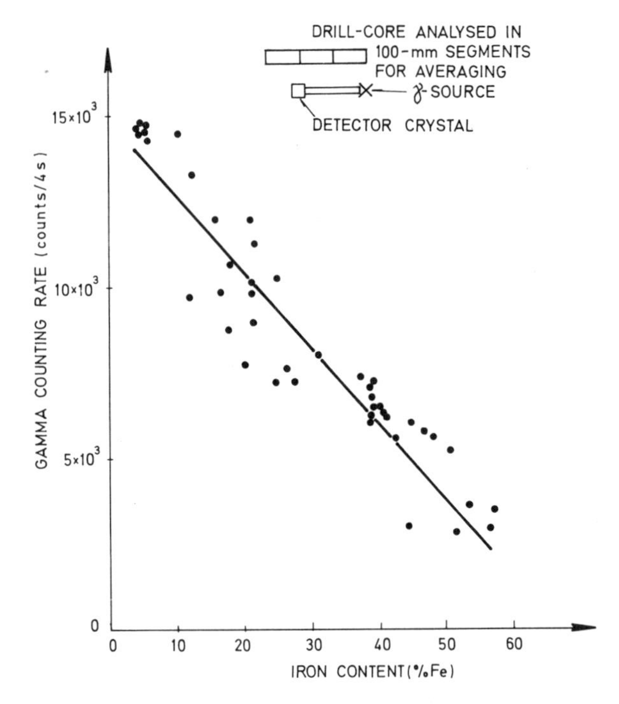

FIG.7. Intensity of back-scattered gamma radiation of energy above 125 keV plotted against iron content in core (γ-γ logging with ^{60}Co at Stråssa).

as well as other elements by neutron activation techniques. A 10-Ci ^{210}Po-Be source (emitting about 2×10^7 n/s) was used for the neutron irradiations. The activation method was tested both in 46-mm diamond-core drilled holes and 250-mm production holes. The results of the borehole measurements were compared with the core analyses.

The ore, which is of the porphyry copper type, is low-grade with an average copper content of about 0.5%.

For technical and economical reasons it is important to feed the dressing plant with ore of approximately constant copper concentration. Thus, if one knows accurately the variation of the copper content of the ore, it is possible to mix low- and high-grade parts of the ore to obtain a constant concentration. Production holes with a diameter of 250 mm were drilled to a depth of 20 m and mud samples from the drillings were collected, transported to the laboratory and analysed for copper. The holes were usually filled with water.

There are, however, some drawbacks connected with the analysis of mud samples, e.g. the difficulty to get representative mud samples and the necessity of transporting the samples to the laboratory for analysis. In-situ analysis by nuclear methods would therefore be an attractive alternative technique. In particular, the vertical distribution of the ore minerals could be accurately determined. As there is a delay of about two months between the drilling and the blasting of the holes there is sufficient time to carry out the analyses.

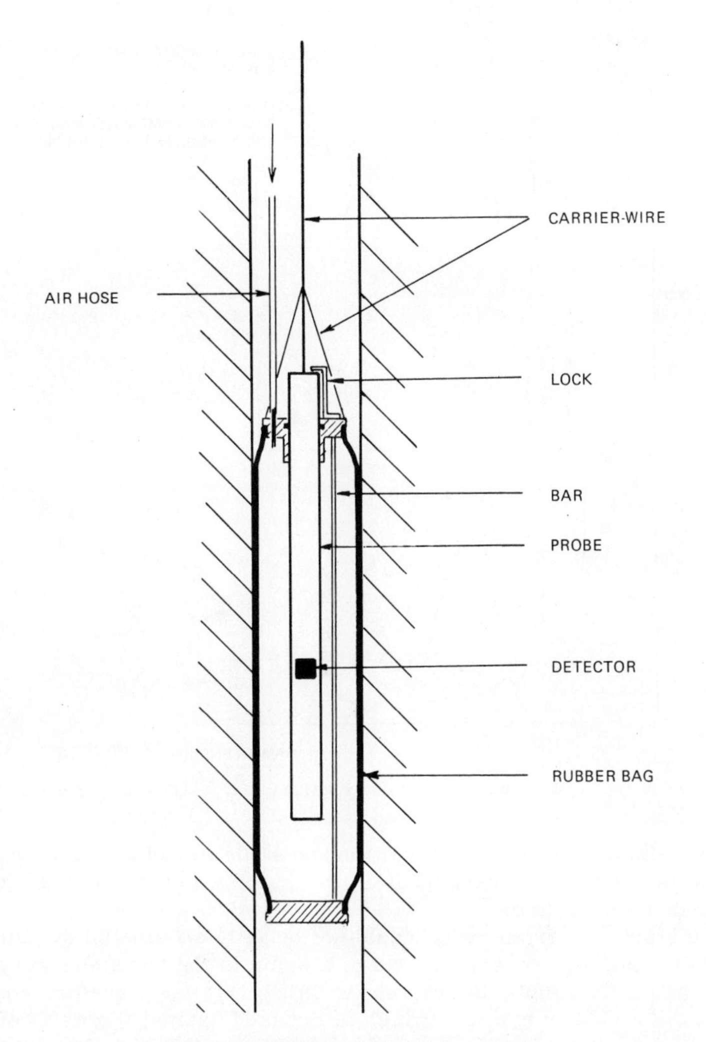

FIG. 8. Logging equipment with rubber device.

Preliminary investigations based on laboratory analyses of rock samples and simulated loggings in borehole models. showed that activation analysis based on the measurement of induced ^{64}Cu activity (0.51 MeV γ, $T_{1/2}$ 12.8 h) would be favourable. Owing to the low copper content in the Aitik ore (~ 0.5% Cu) and the disturbing nuclides ^{28}Al and ^{56}Mn, it would be quite difficult to base the analysis on the short-lived ^{66}Cu (1.04 MeV γ, $T_{1/2}$ 5.1 min).

Activation was done by lowering the neutron source into the borehole where it was kept in the centre of the hole by means of a spring device. After activation – and decay of short-lived activities – measurements were performed with a probe containing a 50-mm-dia. by 50 mm NaI(Tl) crystal.

It was found that the natural gamma radiations in the boreholes were rather intense and that they disturbed the measurements of the induced

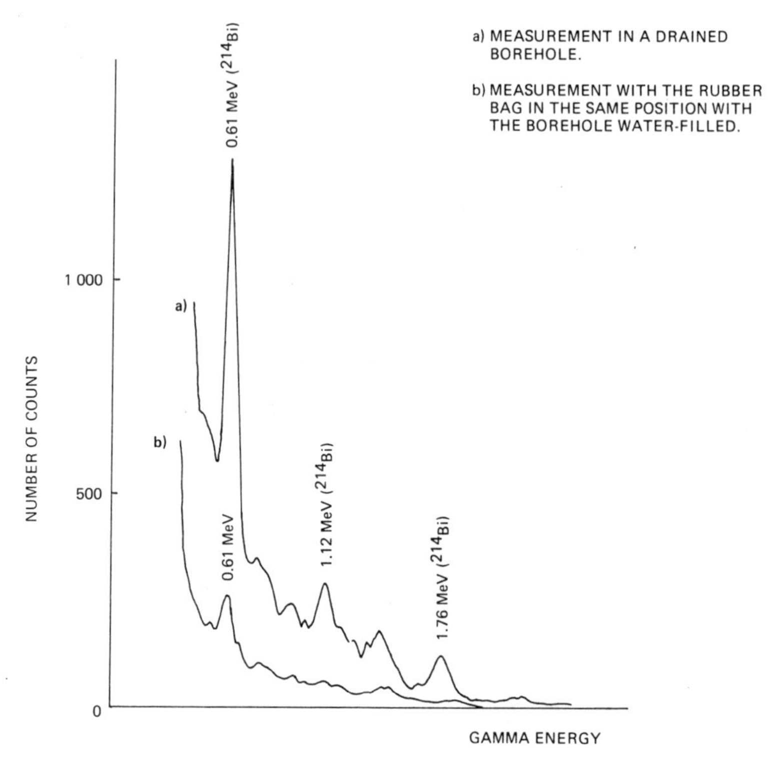

FIG. 9. Test of the efficiency of the rubber device in decreasing radon disturbance.

gamma radiations. The high background was due to radon daughter products present in the boreholes, especially in the water which contained radon in the order of 10^{-8} Ci/litre.

The disturbance from radon could not be fully eliminated by pumping the borehole and, moreover, the use of a pump in the borehole was not very practical. A simple device, shown in Fig. 8, was therefore constructed. It consists of a rubber cylinder which is pressed against the walls of the borehole by filling it with air from a supply of compressed air. The detector is fitted inside the cylinder and the measurements can thus be performed in a radon-free atmosphere. The equipment can be moved quickly to new positions by lowering the air pressure, and it is held in position simply by friction against the walls. With the rubber-bag arrangement the disturbing effect of radon is strongly reduced, as can be seen in Fig. 9.

Comparative measurements with the probe equipped with the rubber bag at a neutral position before and immediately after a 3-h exposure to radon water showed no signs of contamination of the rubber with radon daughter products.

In crystalline rocks the dominant contribution to the induced background comes from the elements aluminium, sodium and manganese; Fig. 10 shows spectra from their nuclides ^{28}Al, ^{24}Na and ^{56}Mn. These spectra were recorded in 250-mm boreholes, but ^{28}Al and ^{56}Mn were also quite easily detected in the diamond-core drilled boreholes (46 mm).

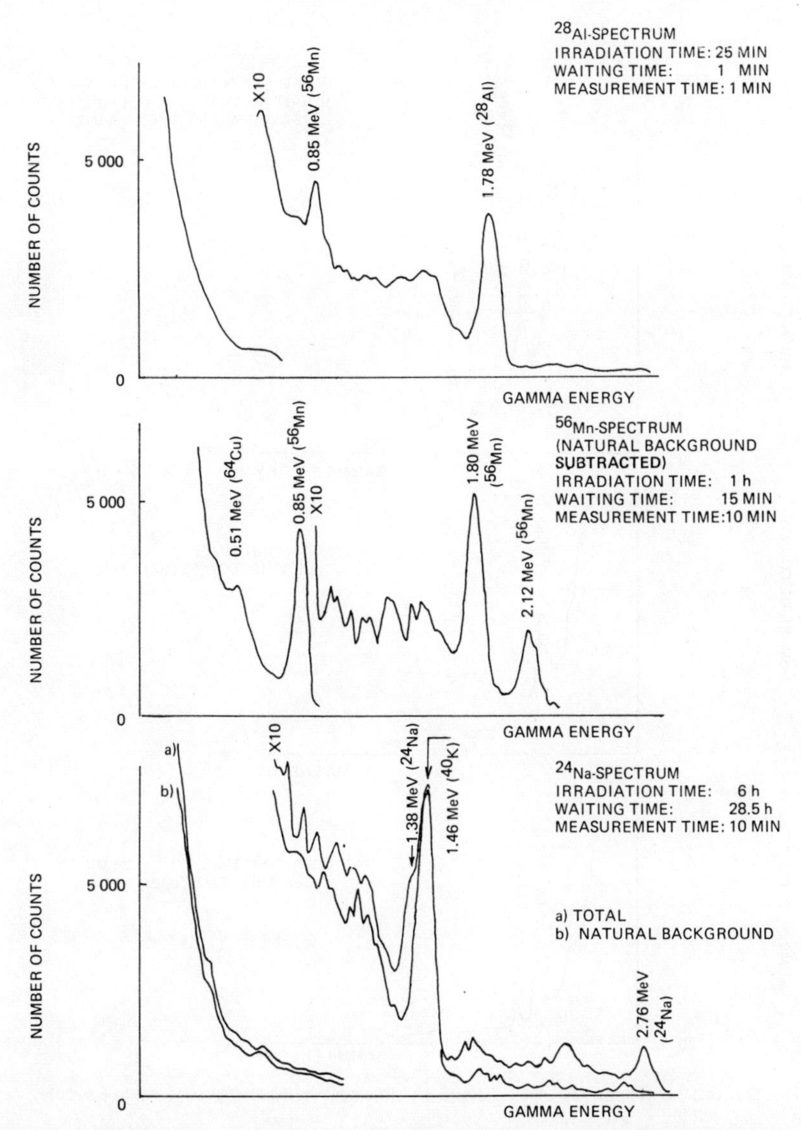

FIG.10. Typical gamma-ray spectra of induced background elements in 250-mm boreholes (neutron activation logging at Aitik).

Some results from the irradiations in an especially copper-rich zone are shown in Fig. 11. The spectra were obtained from a 250-mm hole after an irradiation time of 15 min. Successive measurements were performed in order to study the decay of the nuclides.

The two peaks 0.51 and 1.04 MeV from ^{64}Cu and ^{66}Cu, respectively, can easily be identified in the spectra. The intensities of the two peaks are about the same after an irradiation period of 15 min. A higher intensity of the longer-lived ^{64}Cu is, of course, obtained when a longer irradiation time is used.

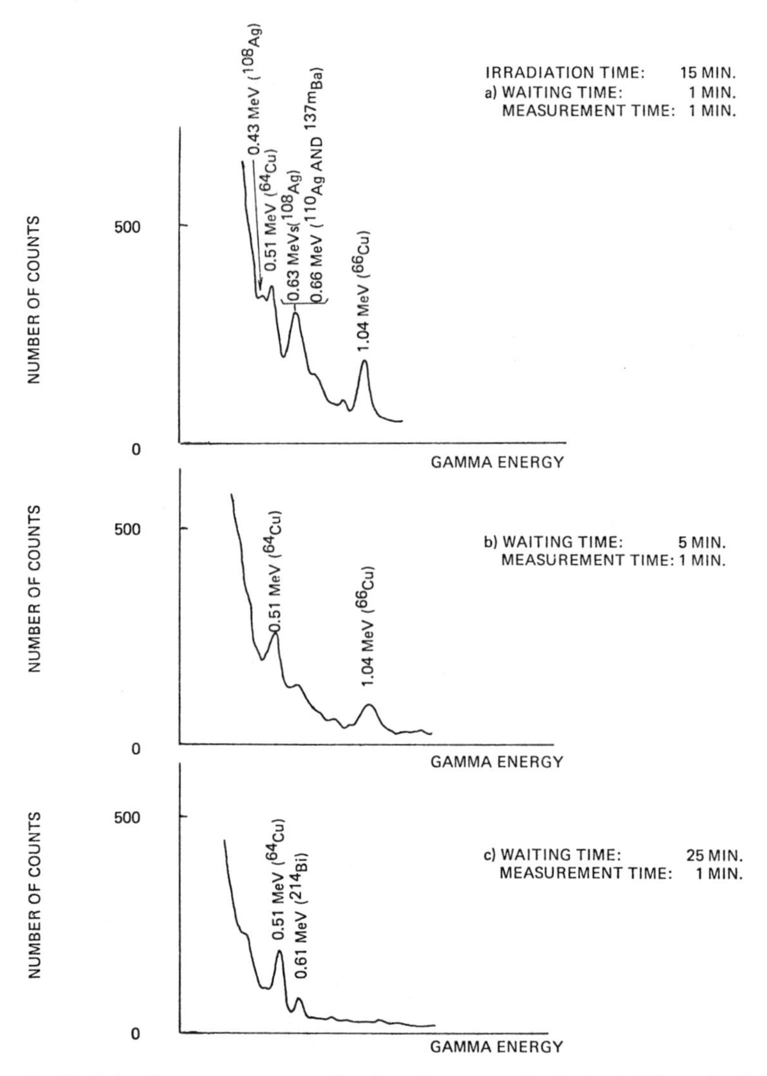

FIG.11. Induced gamma-ray spectra after short-time neutron irradiation in 250-mm borehole.

The broad peak in the region 0.63 - 0.66 MeV probably contains the following gamma-ray energies: 0.63 MeV from 2.4-min ^{108}Ag, 0.66 MeV from 24-s ^{110}Ag and 0.66 MeV from 2.3-min ^{137m}Ba. Of these, the gamma rays from the short-lived ^{110}Ag are noticeable only in the first measured spectrum.

Of the many different nuclides in the Aitik ore that produce photo-peaks around the energy of 0.5 MeV, there is generally none present in sufficient concentrations to seriously interfere with the ^{64}Cu radiation. However, 0.48 MeV from ^{187}W cannot be resolved from the 0.51 MeV of ^{64}Cu by a scintillation detector, and this could create a problem as 0.2% tungsten

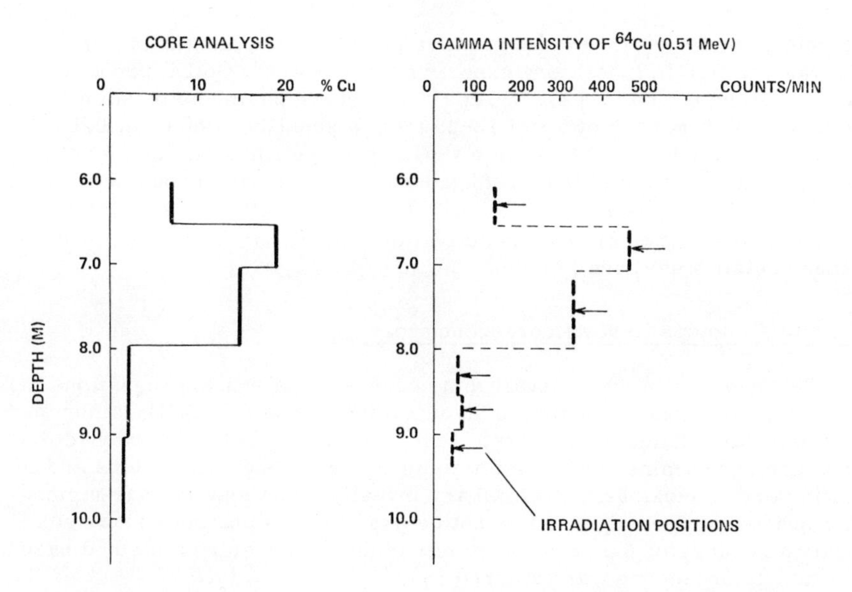

FIG.12. Comparison between core analyses and borehole measurements of induced copper activity (0.51 MeV from ^{64}Cu) after neutron irradiation in 250-mm borehole in a copper-rich zone.

has been found in certain samples of the ore. The 0.51-MeV annihilation gamma rays due to higher energies originate mostly from ^{24}Na and from the high-energy gamma radiation from thorium and uranium, and could be corrected for by normalizing the spectrum.

Figure 12 compares the values obtained from the chemical analysis of copper in the core and the intensity of the 0.51-MeV radiation from ^{64}Cu at the corresponding points of a particular borehole in a very copper-rich zone. The irradiations and measurements were performed separately; during the measurements the rubber device was used. The agreement is fairly good, indicating that the method could be quantitatively calibrated by comparing the measured intensities with the results from core analyses.

To get the desired high sensitivity of the method (< 0.5% copper) a stronger neutron source is needed and its neutron output should be at least of the order of 10^8 n/s. Even if a larger and accordingly more sensitive detector could be used with a weak neutron source in the large boreholes (250 mm), the main reasons for a strong source are to reduce the influence of the natural background and to shorten the irradiation time.

Besides copper, many other elements having high cross-sections and suitable gamma-ray energies of the induced nuclides could be determined in their ores by instrumental in-situ analysis, e.g. gold, manganese, rare earths, silver, tungsten and vanadium. By using an isotopic neutron source the analyses can also be performed in quite small diameter boreholes as has been shown in the present work. This would decrease the cost of the drilling programme.

As all chemical separations are excluded in activation analysis in boreholes, the number of elements that can be determined strongly depends on the resolution of the detector. The use of a Ge(Li) detector for borehole analysis should therefore be of great value in many applications. In fact, this was verified in laboratory experiments in which a Ge(Li) probe was used in borehole models [2]. With a ^{252}Cf source having an output of 10^8 n/s and an irradiation time of about 15-25 min, a sensitivity of about 0.1% Cu should be obtainable. The use of a Ge(Li) probe eliminates interference from other elements and from background radiation, as encountered in the Aitik ore.

The investigation of neutron activation logging has been described in more detail elsewhere [1].

2.4. Radioisotope X-ray fluorescence logging

The lead ore in the Laisvall mine occurs as galena impregnations in quartzitic sandstone belonging to an autochthonous series of Eocambrian and Cambrian sedimentary rocks. Direct assay of lead in the production boreholes in the mine would assist significantly in ore calculations and in locating ore boundaries. Preliminary investigations showed that gamma back-scattering techniques could not be used for unambiguous lead determination because of the occurrence of barium. Therefore a method based on X-ray fluorescence was explored [4].

The K X-rays of lead at 75 keV are excited by means of a gamma radiation source. The corresponding radiation energy for barium is 32 keV. It was decided to register the X-radiation by means of a gamma-ray spectrometer in such a way that each line would fall in its individual energy channel. Two more recording channels were adjusted so as to register the radiation intensity just above each peak to permit matrix corrections to be made by using the ratio of peak channel to adjacent channel count. The principle is demonstrated in Fig. 13. The intensity ratio (called lead and barium ratio, respectively), which is independent of counting periods and source decay, is used as a preliminary measure of the concentration of the metal concerned.

When barium occurs together with lead in the rock matrix, the preliminary approximate concentration value for lead will have to be corrected for the interference from barium, which attenuates the lead X-radiation. This is particularly important for low lead and high barium contents. By means of measurements in borehole models and core-analysed holes it has been possible to determine a correction factor for the lead content for varying barium contents. Figure 14 exemplifies the results from the logging of core-analysed boreholes and shows the effect of the correction procedure for barium.

Calibration of the logging equipment involves fundamental difficulties since the in-situ measurement and the core analyses cannot be performed on the same volume of material. The difficulty is further increased by the heterogeneous composition of the rock material. An attempt to overcome this problem was made by drilling a number of calibration holes in the mine, arranged as in Fig. 15. Boreholes with and without cores (diamond drill-holes and percussion boreholes) were used. Calibration diagrams for the diamond drill-holes are given for lead (Fig. 16) and barium (Fig. 17). The calibration curve for lead is somewhat uncertain at high lead concentrations mainly because of the insufficient number of comparison values. However, this is not very important since an approximate value for the lead concentration is adequate at high concentrations. For low concentrations, e.g. when

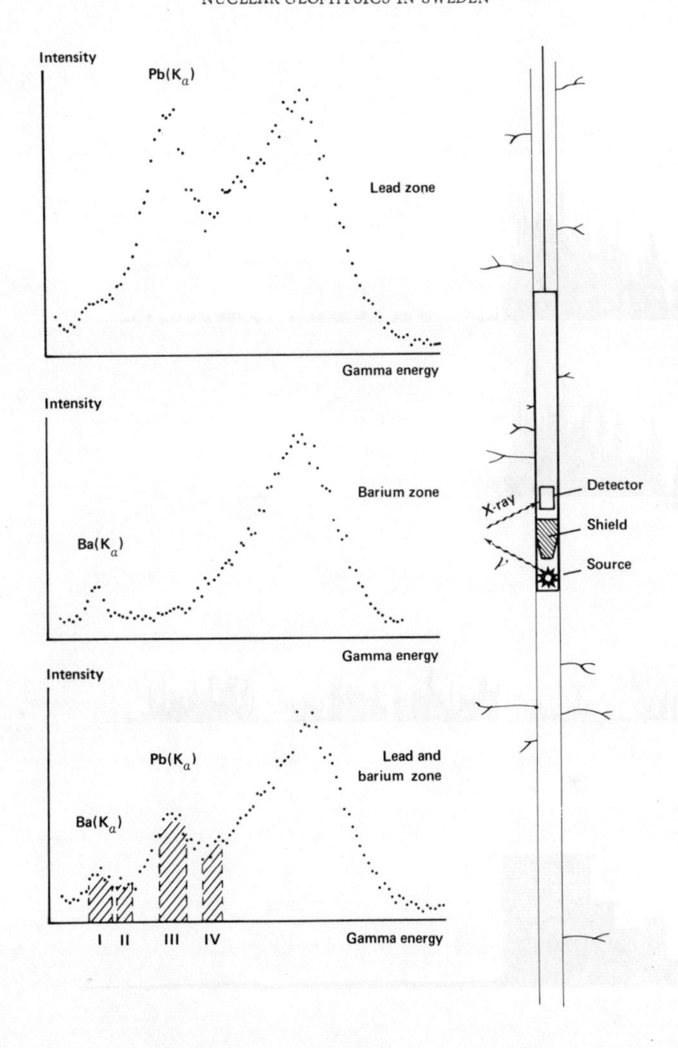

FIG.13. Source/detector configuration and typical spectra from X-ray fluorescent logging (XRF).

determining mining boundaries, the values must be as accurate as possible. The standard deviation of an individual observation, as defined by the least-squares deviation from the best straight line, corresponds to a change in lead concentration of 0.25% Pb.

On the assumption that the calibration curves for the diamond drill-holes also apply for the coreless production holes, a number of holes of each kind were logged and the lead concentrations calculated. These concentrations were then compared with the corresponding results from core analyses. A list of some of these measurements and analyses is given in Table II. The discrepancies noted are not greater than can be explained by the heterogeneous mineralization. From this it can be concluded that the concentration values

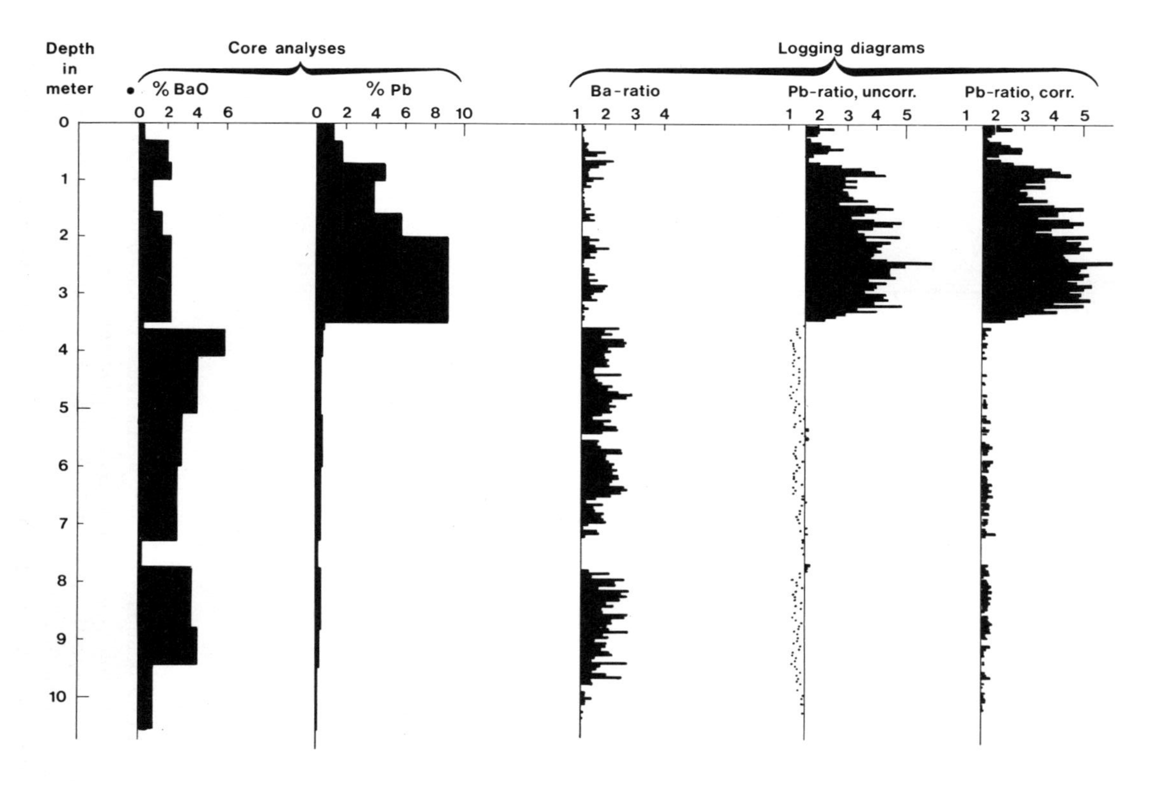

FIG.14. XRF-logging diagrams and core analyses from a calibration borehole in a lead ore (Laisvall). The effect of correction of the Pb ratio for the influence of barium is shown.

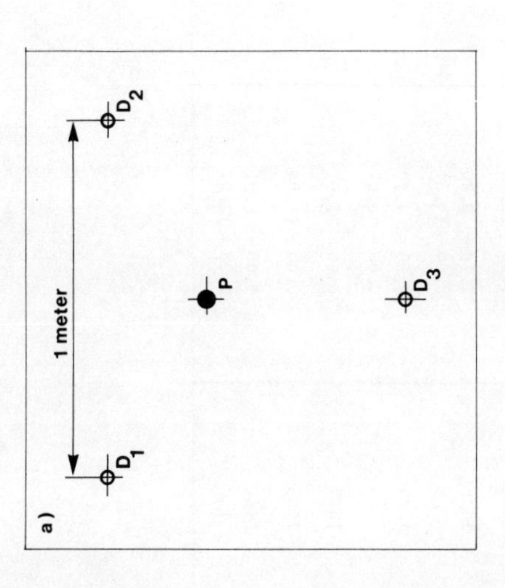

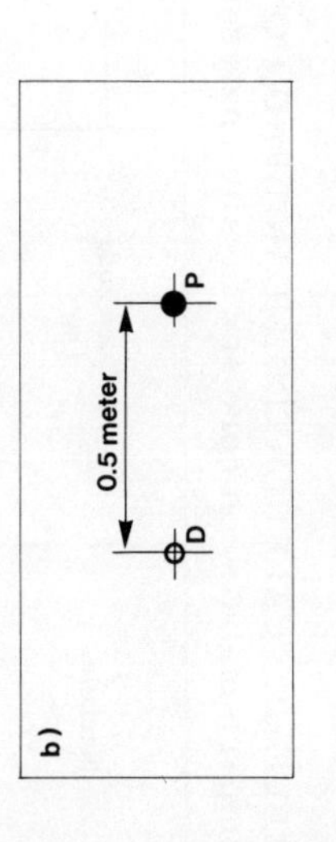

FIG.15. Geometric pattern of (a) the lead-barium calibration boreholes and (b) the holes for the accuracy test.
D = Diamond (core) drilled 46-mm holes. P = Coreless 51-mm holes.

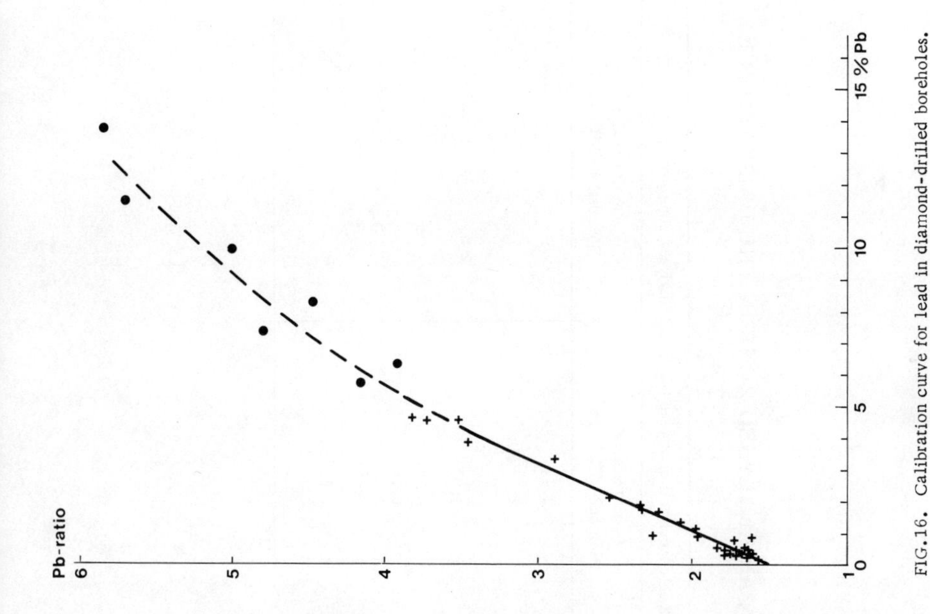

FIG.16. Calibration curve for lead in diamond-drilled boreholes.

TABLE II. CALCULATED MEAN LEAD CONCENTRATION IN AN ANALYSED ZONE IN THE LAISVALL MINE
Comparison between core and in-situ analyses

Borehole	Analysed zone (m)	Calculated mean concentration of lead (%) in the analysed zone from:		
		Core analysis	In-situ analysis Diamond-drilled hole	In-situ analysis Percussion-drilled hole
1435	0.00 - 11.48	2.56	2.44	2.10
1436	7.10 - 9.86	2.52	2.31	
	0.00 - 11.54	1.20		1.29
1437	0.00 - 4.79	1.78	1.93	
	0.00 - 6.07	2.16		2.31
1438	0.00 - 11.01	7.18	6.76	6.03
1439	0.00 - 11.04	0.45	0.46	0.46
1440	0.00 - 8.91	0.36	0.42	
	0.00 - 12.07	1.12		1.33

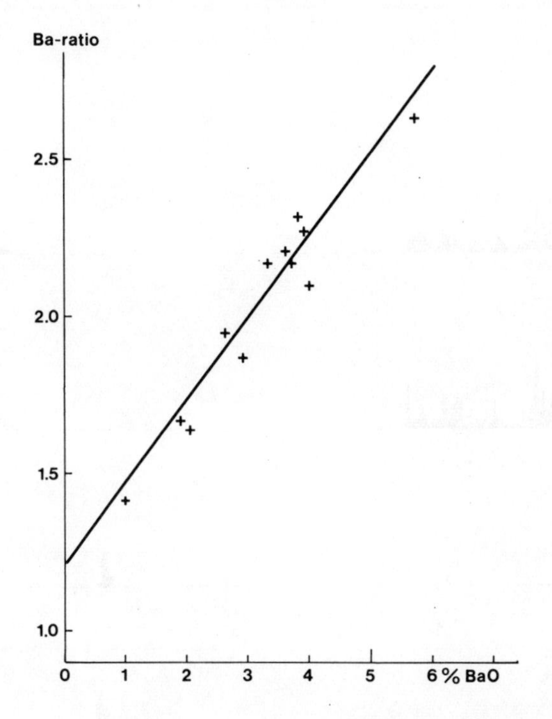

FIG.17. Calibration line for barium in diamond-drilled boreholes.

for lead obtained by the radioisotope X-ray fluorescence method are as reliable as those obtained by core-drilling and analysis. Furthermore, the logging technique is much quicker and it enables a very detailed investigation of the borehole profile to be made even in coreless holes, as exemplified by the logging diagram in Fig. 18 which reveals that all the barium is concentrated in one rich barite zone at 8.5 m. The use of coreless drilling instead of the more expensive diamond-core drilling would also make considerable savings in drilling costs while retaining or improving the information on metal concentrations.

2.5. Other techniques

Nuclear activation analysis has been used in a large number of instances for laboratory analysis of geological samples of widely varying origin by AB Atomenergi and by ITL.

At AB Atomenergi a spectrometric version of the method in combination with an automated, rapid chemical group separation is used [5, 6], and at ITL a number of elements (Hg, As, Se, Cd, Ag, Au and Co) are determined using complete chemical separation.

The attempts to use the gamma radiation emitted upon thermal neutron capture for geophysical analyses of iron ores have not been successful due to saturation effects which make the method insensitive to concentration changes

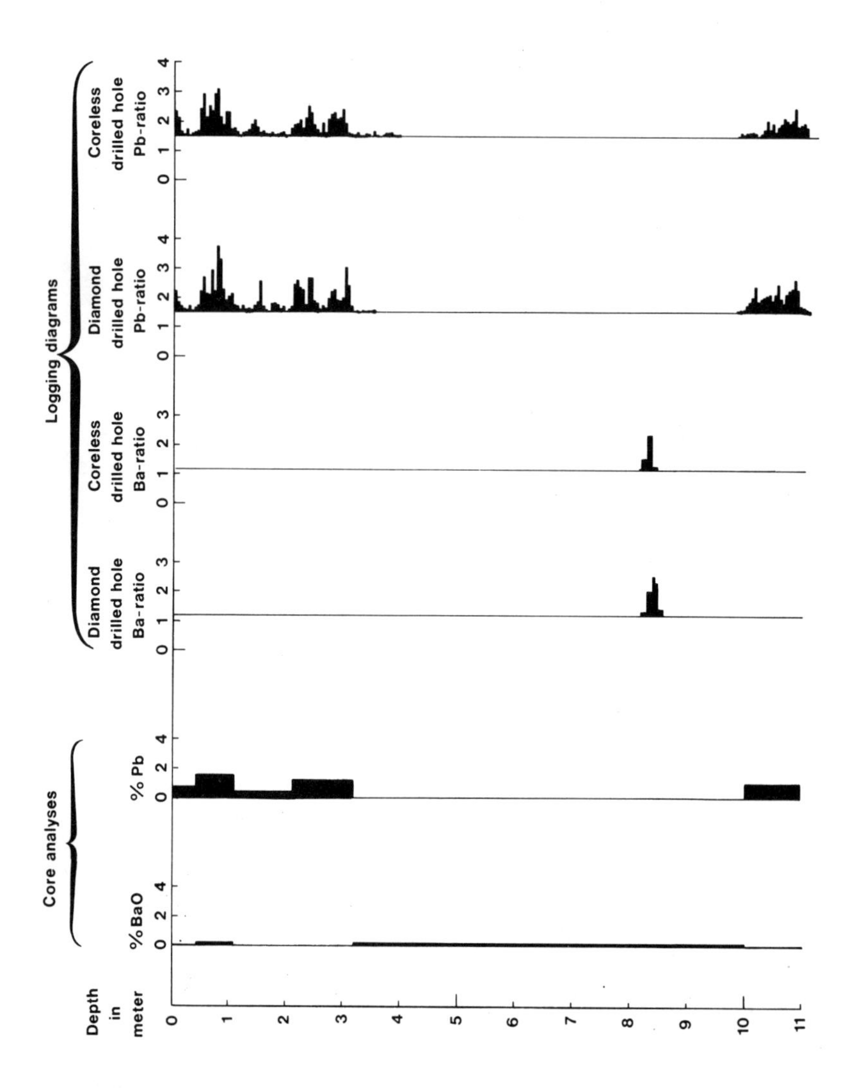

FIG.18. Logging diagrams and core analyses from a low-grade lead zone in the Laisvall mine.

above some 20 to 30% of iron. This is too low for Swedish ores. On the other hand, the method works well for iron determination on belt-conveyed ores and ore concentrates where the total amount of iron seen by the source/detector arrangement is limited. The method is described in these Panel Proceedings [7].

REFERENCES

[1] LANDSTRÖM, O., CHRISTELL, R., KOSKI, K., Geoexploration 10 (1972) 23.

[2] LAUBER, A., LANDSTRÖM, O., Geophys. Prospecting 20 (1972) 800.

[3] LANDSTRÖM, O., MALMQVIST, D., Correlation between radioactivity content of rocks and heat flow in some Swedish mining districts, to be published.

[4] LANDSTRÖM, O., et al., Quantitative in-situ determination of lead and barium in boreholes using a radioisotope X-ray fluorescence logging technique, to be published.

[5] LANDSTRÖM, O., WENNER, C.G., Rep. AE-204, AB Atomenergi, Stockholm (1965).

[6] LANDSTRÖM, O., SAMSAHL, K., WENNER, C.G., Rep. AE-296, AB Atomenergi, Stockholm (1967).

[7] LJUNGGREN, K., CHRISTELL, R., On-line determination of the iron content of ores, ore products and wastes by means of neutron capture gamma-radiation measurements, these Proceedings.

ANALYSIS OF ELEMENTS IN BOREHOLES BY MEANS OF NATURALLY OCCURRING X-RAY FLUORESCENCE RADIATION

O. LANDSTRÖM
AB Atomenergi,
Studsvik, Nyköping,
Sweden

Abstract

ANALYSIS OF ELEMENTS IN BOREHOLES BY MEANS OF NATURALLY OCCURRING X-RAY FLUORESCENCE RADIATION.

A technique is described which utilizes naturally occurring X-ray fluorescence radiation to analyse heavy elements in boreholes. Thus, no artifical radiation source is needed for the X-ray excitation. In preliminary experiments, lead and barium minerals were identified. The K X-ray intensity of lead was correlated to the core analyses of lead and it is shown that the technique can be used for a qualitative identification of an ore and for locating its boundaries. The technique also has potential for quantitative or semi-quantitative analysis of lead. A drawback with the technique, compared with techniques using artificial radiation sources, is the low intensity of the natural X-ray fluorescence radiation which necessitates fairly long measuring periods.

1. INTRODUCTION

The paper describes some preliminary work on the use of naturally occurring X-ray fluorescence radiation for the analysis of elements in boreholes. The potential of this method is also discussed. The experiments were performed in the Laisvall lead mine in northern Sweden in connection with development work on the application of X-ray fluorescence techniques to the in-situ assay of lead [1].

2. METHOD

Characteristic X-rays from different elements in rocks and ores are excited mainly by the alpha, beta and gamma radiation from different nuclides in the decay chains of thorium and uranium and from the potassium isotope ^{40}K. Owing to the short penetration range of alpha and beta particles in dense material such as rocks and minerals only those atoms which are close to the emitting nuclides are excited by these particles. The greater penetrating power of gamma rays makes them more effective since atoms in adjacent, non-radioactive, minerals can also be excited. (This is true also for the bremsstrahlung produced by beta particles).

The intensity of the X-ray fluorescence radiation depends on the intensity of the natural gamma radiation and thus on the concentration of potassium, thorium and uranium in the rocks. In spite of the relatively low intensity of the natural gamma radiation in "normal" rocks (compared with that from artificial sources for X-ray excitation), measurable intensities of the X-ray

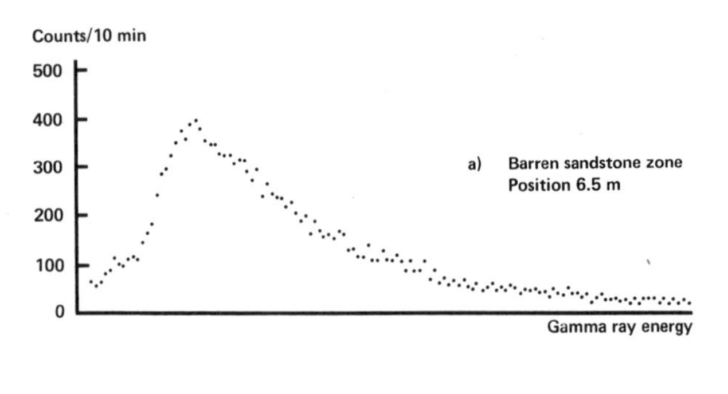

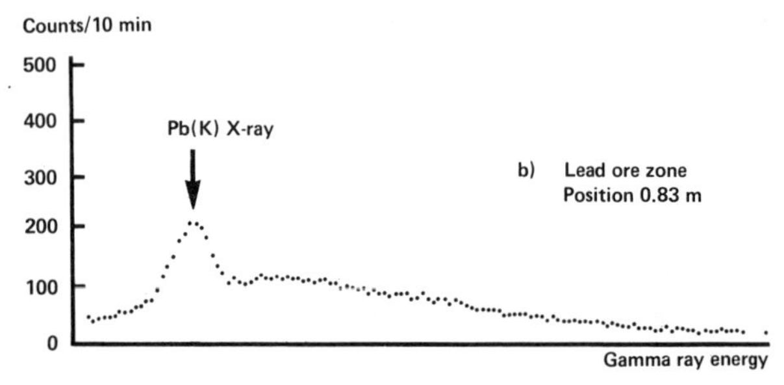

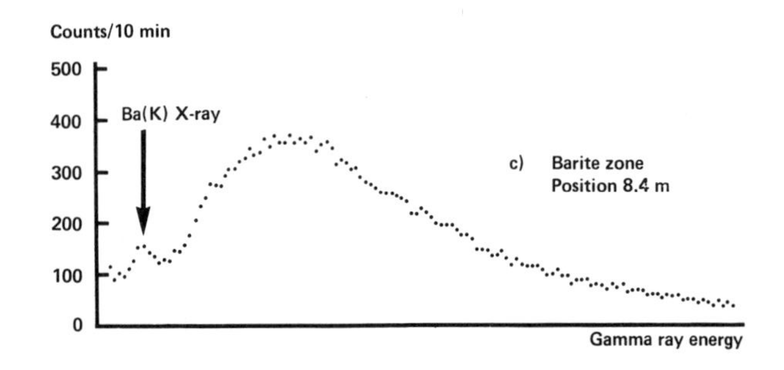

FIG.1. The low energy part of natural spectra measured with a 1 in × $1\frac{1}{4}$ in NaI detector in borehole 1439 (46 mm) in the Laisvall lead mine.

fluorescence radiation are excited if the concentration of the element of interest is unusually high, as is the case for certain ores and minerals. This is exemplified in Fig.1 which shows the low energy region of three natural spectra measured with a 1 in × $1\frac{1}{4}$ in NaI detector in a diamond-drilled borehole in the Laisvall lead mine.

Figure 1(a) is a spectrum from a barren sandstone and Fig.1(b) is the corresponding spectrum from an ore-bearing sandstone in which the characteristic K X-ray emission from lead is observed. In Fig.1(c) the characteristic K X-ray emission from barium can be identified in the spectrum recorded in a barite zone.

3. OUTLINE OF GEOLOGY AND NATURAL RADIOACTIVITY OF THE TEST AREA

The field experiments were carried out in underground boreholes in the Laisvall lead mine situated in the north-western part of Sweden. The main ore mineral is galena which is dispersed in Eocambrian quartzitic sandstones. The average lead concentration of the ore is 4% but the concentration varies so much that in some places there may be as much as 40% lead. Among the associated minerals barite deserves special attention since this mineral can occur in high enough concentrations to disturb the X-ray fluorescence measurements of lead [1].

The ore-bearing sandstones belong to an autochthonous sequence of sedimentary rocks deposited on a Precambrian granite basement. Starting from the bottom, the strongly eroded granite is covered by arcoses, argillaceous shales and the sandstone formation, which is about 40 m thick.

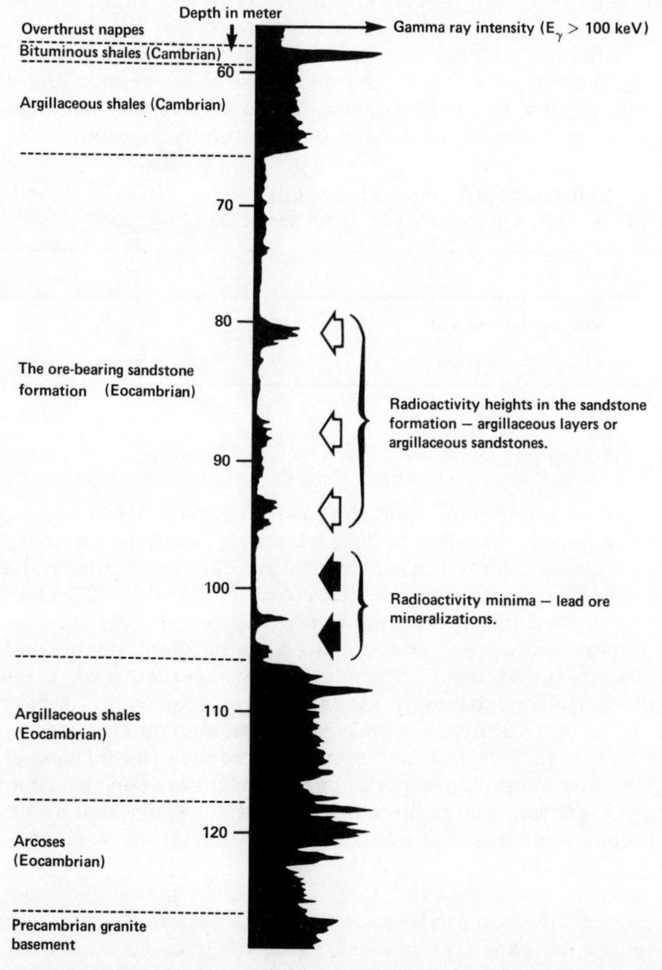

FIG.2. Gamma-ray log of natural radioactivity. Borehole 874 in the Laisvall lead mine.

Above the sandstones are Cambrian sediments covered by overthrusted nappes. More detailed information about the Laisvall ore and the geology of the Laisvall area is given in Refs [2 - 4].

The level of natural radioactivity in the different rocks can be seen in the gamma-ray logging diagram in Fig.2. The radioactivity of the ore-bearing sandstone is very low compared with that of the over- and underlying sedimentary rocks. In fact, the presence of lead ore is easily recognized in logging diagrams by the unusually low level of radioactivity due partly to the strong absorption of gamma rays by lead. Zones of increased radioactivity are, however, exhibited by the sandstone formation. These are argillaceous layers or argillaceous sandstones whose potassium, thorium and uranium concentrations are higher than those of pure sandstone. It is noteworthy that lead mineralization in these argillaceous layers is very low or completely absent [2, 3]. As only a few analyses of potassium, thorium and uranium in ore-bearing sandstones and argillaceous layers have been made, the results shown in Table I should be taken as an indication of the order of magnitude rather than as a representative figure.

Thus, although the lead emission is mainly produced by radioactive elements in the lead ore itself, the presence of corresponding elements in argillaceous layers and argillaceous sandstones is also significant, especially when they occur close to the region of ore mineralization.

TABLE I. CONCENTRATION OF POTASSIUM, THORIUM AND URANIUM IN THE ORE-BEARING SANDSTONE

	K	Th	U
Sandstone with argillaceous shale	2.6%	9.5 ppm	1.4 ppm
Sandstone with lead mineralization	0.51%	0.9 ppm	0.5 ppm

4. EXPERIMENTAL

The logging equipment used in the field experiments has been described in an earlier paper [1]. The probe, which was equipped with a 1 in $\times 1\frac{1}{4}$ in NaI detector, was mounted in an aluminium casing 43 mm in diameter and 1 mm thick at the position of the detector. A portable 128-channel analyser was used to record the spectra shown in Fig.1.

The experiments were concentrated on the diamond-drilled borehole No. 1439 (diameter 46 mm). Stationary measurements were made at a large number of vertical positions in steps of 5 cm, except in the barren sandstone between 3.2 m and 10 m where only seven measurements were made. The sampling time was 2 - 4 min per point. To reduce the influence of matrix effects, measurements were made of the ratio between the count-rates in two energy windows, one at the characteristic X-ray peak and the other at slightly higher energies. The technique is similar to that used in the study described in Ref. [1].

The results are shown in Fig.3 in which the measured lead ratios are compared with the core analyses. The ratio obtained in the barren sandstone was chosen as the base line in the logging diagram (the mean of seven measured values).

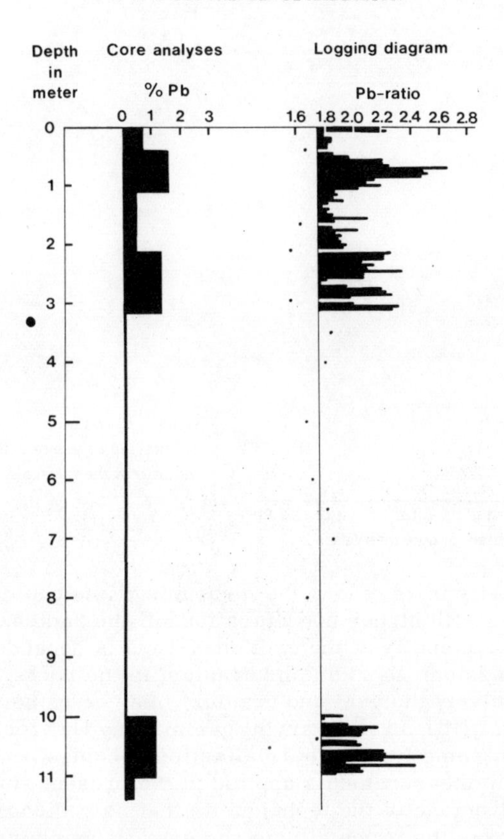

FIG.3. Comparison of the core analyses and the logging diagram of the Pb ratio for borehole 1439 in the Laisvall lead mine.

5. DISCUSSION

The logging diagram in Fig.3 gives a fairly good picture of the extent of the mineralization along the borehole, as verified by visual observation of the core, and the technique can, without doubt, be used for qualitative identification of an ore and for locating its boundaries. The statistical errors can certainly be reduced if longer sampling times are accepted.

The assumption that the technique is capable of accurate quantitative analysis remains to be tested. Some indication of the possibilities is given in Fig.4 where the mean lead content of the analysed core intervals is correlated with their mean lead ratio. In this connection it is necessary to consider the different sources of error involved in a comparison of results of core analyses and borehole measurements. Different materials are being compared, namely the material surrounding the hole and that in the core; the ore minerals may be inhomogeneously distributed, etc. [1].

The main drawback with the technique described here, as compared to techniques using radioactive sources, is the low intensity of the natural X-ray fluorescence radiation which necessitates fairly long measuring periods. In the lead ore, which formed the subject of the present study, the exciting

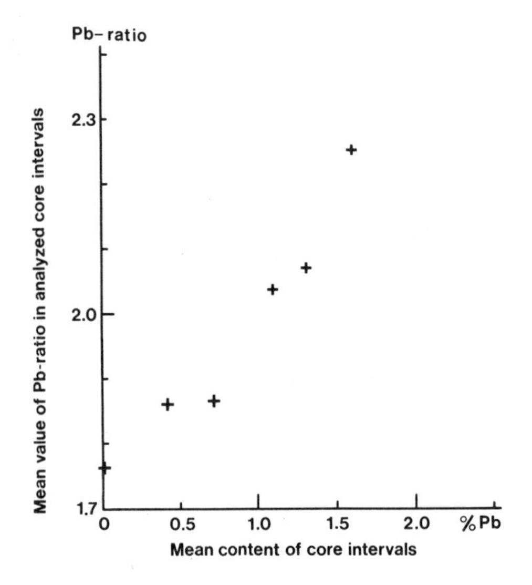

FIG.4. Correlation between the lead concentration of analysed core intervals and the mean value of the PB ratio of the intervals.

natural radioactivity is very low; the measuring time could possibly be decreased in ores with higher concentrations of the radioactive elements. Furthermore, the intensity of the excited X-rays is dependent on the concentration of potassium, thorium and uranium in the rocks. The trace elements, particularly thorium and uranium, may be rather inhomogeneously distributed which will lead to a varying gamma-ray flux for the X-ray excitation. Thus, some form of normalization becomes necessary as, for example, the ratio measurements applied in the present study.

The main advantage of the technique is that no radioactive source is needed. This makes the probe simple and easy to handle and avoids safety problems. Measurements can be performed in boreholes where, for some reason, the introduction of a radioactive source is prohibited. The technique becomes particularly attractive if such a borehole contains mineralized zones from which recovery of cores or representative cuttings is impossible. The technique can be applied to heavy elements other than those tested in the present study, such as tungsten. It can also be readily applied in connection with conventional gamma-spectrometric measurements of the natural activity from potassium, thorium and uranium thus extending these measurements to non-radioactive elements and minerals.

For light element analysis the strong absorption of low energetic X-rays will introduce problems similar to those associated with other methods [1].

REFERENCES

[1] LANDSTRÖM, O., et al, Quantitative in-situ determination of lead and barium in boreholes using a radioisotope X-ray fluorescence logging technique, to be published.

[2] GRIP, E., The lead deposits of the eastern border of Caledonides in Sweden, Int.Geol. Congr., XXI Session, Copenhagen (1960) Pt 16, p. 149.

[3] GRIP, E., On the genesis of the lead ores of the eastern border of the Caledonides in Scandinavia, Econ. Geol., Monograph 3 (1967) 208.

[4] LILLJEQUIST, R., Caledonian geology of the Laisvall area, southern Norrbotten, Swedish Lapland, Sveriges Geol. Undersökn.,Ser. C, No. 691 (1973): (Årsbok 67, Nr 10).

THE USE OF NATURAL GAMMA RADIATION FOR ESTIMATING THE IRON CONTENT OF SEDIMENTARY IRON FORMATIONS CONTAINING SHALE BANDS

J. A. AYLMER, P.L. EISLER, P. J. MATHEW,
A.W. WYLIE
CSIRO,
Port Melbourne,
Victoria, Australia

Abstract

THE USE OF NATURAL GAMMA RADIATION FOR ESTIMATING THE IRON CONTENT OF SEDIMENTARY IRON FORMATIONS CONTAINING SHALE BANDS.

Natural gamma radiation has been logged to estimate the mean grade of iron ore in dry boreholes traversing layered mixtures of hematite and shale. The shale contains uranium and thorium varying in abundance from place to place about a mean value. The field data was subjected to regression analysis to determine the mean grade of hematite in single drill holes. The pooled, transformed, data from all holes was then utilized to derive the final expression for prediction of <u>mean</u> ore grade from the mean count-rate of a given hole, and the 95% confidence limits of ± 0.8% Fe for this mean grade. The method is applicable to both exploration holes and blast holes.

1. INTRODUCTION

The lower Proterozoic sediments of the Hamersley Iron Province of north-western Western Australia contain several banded iron formations, all of which are unmetamorphosed and only moderately folded. Enrichment of these, particularly enrichment of the Brockman Iron Formation, has produced several large deposits of high-grade hematite by processes in which siliceous materials and carbonates have been removed from the original formation concurrently with mobilisation and redeposition of iron at lower levels [1, 2].

Numerous shale bands, also enriched in iron, are interbedded with the enriched hematite zones. Distributed within these bands, which vary in thickness from a few centimetres to several metres, are approximately 10 - 20 ppm of both uranium and thorium. Potassium may or may not be present. At the Tom Price iron ore deposit investigated in this work it is present in the predominantly kaolinitic shale bands traversing hematite in amounts of approximately 0.05% or less (cf. Section 4.2).

It is the universal presence of the natural radioactivity of uranium and thorium in these shale bands that enables them to be readily identified and the stratigraphy of the area to be soundly established by qualitative observations, not only in the neighbourhood of working mines as demonstrated by Jones and coworkers [3], but in principle ultimately over the whole of the iron ore province.

Quantitative measurements of natural gamma-ray activity, however, have not previously been made with a view to establishing the grade of iron ore in a geological section such as a drill hole. This paper reports on the results of field investigations carried out in 1972-1973 largely at the Mt. Tom Price Mine of Hamersley Iron Pty. Ltd. The paper demonstrates that the mean iron grade in a geological section such as a blast or other drill hole can be established to within 95% confidence limits ± 0.8% Fe. Thus, both stratigraphy and mean ore grade for the section can be established simultaneously. Prediction accuracy for estimation of grade from measurement of radioactivity of single split-samples is too low to be useful, mainly owing to variability in the distribution of natural radioactivity.

2. PHYSICAL BASIS OF METHOD

The model assumed for the proposed technique is that of one or more shale bands, carrying uniformly admixed or adsorbed natural gamma-ray activity, interbedded with hematite bands carrying relatively little natural radioactivity. Each shale band contains hematite uniformly distributed throughout and each hematite band contains either microbands of shale or small, more or less uniformly distributed pockets of shale. By estimating the shale content of the overall mixture from the observed radioactivity, the hematite fraction of the two-component system just described can be obtained and the iron grade calculated.

The basic logging theory for a simple "one-layer" model of the above type has been reviewed by Czubek [4] who presents the following expression for the intensity, I, of unscattered gamma radiation at points within an empty borehole crossing one radioactive layer of thickness t and infinite lateral extension:

$$I = 4\pi k \frac{\rho}{\mu} Q \qquad (1)$$

where Q is the radioactivity per gram of rock, ρ/μ is the reciprocal of the mass attenuation coefficient for radiation of energy used in assessing Q, and k is a constant.

This indicates that the radiation intensity within the borehole due to an infinite, homogeneous matrix is independent of borehole diameter and excentricity of the probe[1]. Furthermore, I is determined solely by Q only if ρ/μ is constant. With mixtures of hematite and shale, however, there is a progressive increase in ρ/μ, the reciprocal of the mass attenuation coefficient, as the percentage of shale is increased. For example, ρ/μ, decreases at 1.5 MeV from 20.2 to 19.1 g/cm^2 if the material changes from pure hematite to pure shale. Thus, I is a function of both Q and ρ/μ and, in the absence of obscuring factors, a significant departure from a linear response between I and Q must be expected because of the more effective electronic screening of the low Z materials compared to hematite.

---

[1] This was verified by the authors in a series of ilmenite beach sand models furnished with various paper tubes simulating boreholes of different diameter. This sand has greater radioactivity and higher density than alternative materials available for such experiments.

For a multi-component rock, the attenuation coefficient is given by

$$\mu = \rho N_0 \sigma_e \sum_{i=1}^{i=n} \frac{Z_i}{A_i} P_i$$

where N_0 is the Avogadro number, σ_e is the Compton scattering cross-section for the electron, and Z_i, A_i, P_i are respectively the atomic number, atomic weight and fractional concentration of the i-th constituent element.

For a hematite-shale system this can be simplified to:

$$\mu = (\rho N_0 \sigma_e) \{(\sum_s Z_s/M_s) C_s + (1-C_s)(\sum_H Z_H/M_H)\} \quad (2)$$

where M_s and M_H are the molecular weights of shale and hematite, respectively, and C_s and C_H are the fractional concentration of shale and hematite respectively.

Substitution of this expression into the following differential equation

$$\frac{dI_p}{dC_s} = \frac{\partial I_p}{\partial Q} \frac{dQ}{dC_s} + \frac{\partial I_p}{\partial(\rho/\mu)} \frac{d(\rho/\mu)}{dC_s} \quad (3)$$

where I_p is the intensity of the primary radiation, leads to the following approximate results:

$$\frac{dI_p}{dC_s} \simeq K \{1 - 2 xC_s + 2(xC_s)^2\} \quad (4a)$$

where

$$x = \{(\sum_s Z_s/M_s) - (\sum_H Z_H/M_H)\}/(\sum_H Z_H/M_H)$$

$$\simeq 0.058$$

$$K = 4\pi \, kq_s/N_0 \sigma_e \sum_H Z_H/M_H \quad (4b)$$

where $\sum_s$ and $\sum_H$ represent the summations for all constituents of shale and hematite respectively and q_s is the (assumed) constant specific activity of shale.

Thus, the incremental loss from linearity of dI_p/dC_s is approximately $0.116\ C_s$.

Again the intensity I_{scat} of the scattered radiation, according to Czubek [4], is proportional to I_p so that:

$$\frac{dI_{scat}}{dC_s} \propto \frac{dI_p}{dC_s}$$

although the measurements to be described were carried out in an energy window with a low-energy threshold of 400 keV and an upper limit of 3.0 MeV, that is, in a region where the predominant photon collision process is Compton scattering. The shape of the observed spectrum therefore remained essentially independent of chemical composition. However, photoelectric and pair production processes cannot be totally ignored in the above energy range and their effect for $E_P > 1$ MeV, according to the curves predicted on theoretical grounds by Goldstein and Wilkins [5] for constant mean-free-path thickness of material, indicate that there will be some further suppression of spectral intensity above ~400 keV for low Z materials relative to high Z materials.

Again, additional suppression of spectral intensity in the above manner can also arise in a borehole system because the average number of mean free paths traversed per photon detected will increase as the concentration of shale increases relative to hematite. Thus, if

$$I' = I_p/4\,kQ$$

and

$$\rho/\mu \simeq M_H/N_0 Z_H \sigma_e \quad (1 - xC_s)$$

then

$$\frac{dI'}{I'} \simeq -x(1 - xC_s - x^2C_s^2)\,dC_s \qquad (5)$$

$$\simeq -x\,dC_s$$

which signifies an overall decrease in the transmission-cum-detection efficiency of the detector-matrix system as the shale content increases. Hence, the average number of collisions experienced per photon detected increase in the same way as I' decreases with increasing shale content, causing the spectral redistribution referred to.

In summary, then, these three effects would be expected to cause an increasingly greater deviation from linearity of the detected intensity as shale content increases when applied to integral counts above some threshold energy such as 400 keV than when applied to counts under a single photopeak, for example that due to ^{208}Tl (ThC'' 2.62 MeV).

In practice, as will be shown later, the variation of the specific activity of various shales encountered has largely been found to obscure these systematic but relatively small departures from linearity of the I versus C_s relationship.

3. LABORATORY AND FIELD WORK

It is difficult to prepare realistic models of shale and iron ore in the laboratory to test adequacy of a postulated linearity in response, particularly with very large drill holes such as blast holes, and investigations were therefore carried out using assay data from drill holes as described later. However, a preliminary laboratory investigation was first carried out with one simple model consisting of a mixture of shale and iron ore to establish the nature of the probe response in a smaller, exploration-type hole. Parallel with this experiment, laboratory and field investigations were begun to test both the chemical and the geological variability of radioactivity in different sections of the deposit.

3.1. Model borehole responses

The simplified "one-layer" laboratory model consisted of a mixture of shale and hematite in different proportions arranged around thin-walled paper tubes of 13-cm bore, axially located in a series of plastic drums of 50-cm depth. The width of the drums was selected so that the lateral dimensions of the layer were such as to contain four mean-free-paths of the highest energy photons, although this requirement was not quite met in the case of the pure shale model. The iron ore contained 63% Fe and some 10% residual shale from the mining operation. The added shale was from a local Melbourne quarry. Chemical assay showed that it contained potassium (1.5% K) as well as thorium (10 ppm) and uranium (4 ppm). Weighed quantities of the two components were mixed and ground to pass completely through a fine screen (opening 0.35 mm) before further intimate mixing in a blending machine, pouring into the models, and consolidation to equalise volumes. It was found that blending times should be short to avoid segregation of the components.

A logging probe housed in an anodised aluminium tube and containing a 5 cm × 5 cm NaI(Tl) scintillator was used to obtain the responses in various spectral regions: total response, 400 keV – 2.71 MeV; potassium response, 1.36 - 1.56 MeV; uranium response, 1.66 - 1.81 MeV; and thorium response, 2.44 - 2.71 MeV. The probe was connected to a 400-channel pulse height analyser calibrated with ^{137}Cs and ^{60}Co sources. The lifetime was 300 s. Gain was stabilized during this period with an electronic gain stabilizer set on the peak of thorium at 2.62 MeV. Background subtraction to give net count-rates was made by use of the programme developed for control of the digital logging operation. Readings were repeated five times and averaged to give the results presented in Fig. 1. This procedure was adopted to overcome drift problems caused by lessened ability of the gain stabilizer to work at its optimum capacity at low count-rates.

A typical borehole spectrum of mine-area shale is shown in Fig. 2. The absence of a potassium peak in this and in all other mine-area shale spectra[2] distinguishes field spectra from spectra of the local shale used for convenience in the laboratory tests.

3.2. Field work

Field work was carried out with similar probe equipment, now under control of a PDP8/L mini-computer with 8 K memory capacity. This digital borehole logging system [6, 7] is linked to the winching and depth-measuring gear and furnishes the average net count-rate for any selected spectral window at successive depth intervals, set in the present instance at 61 cm (2 ft) as the probe travels up or down the borehole. These intervals correspond to the 61 cm "splits" recovered by the percussion drilling crew from the all-dry holes in the form of dust samples for later examination and chemical assay. The logging speed was 61 cm/min (a larger crystal and faster electronics would enable this to be increased several times) and sampling time was ~ 40 seconds live-time. Results were available on

[2] Footwall shales excluded.

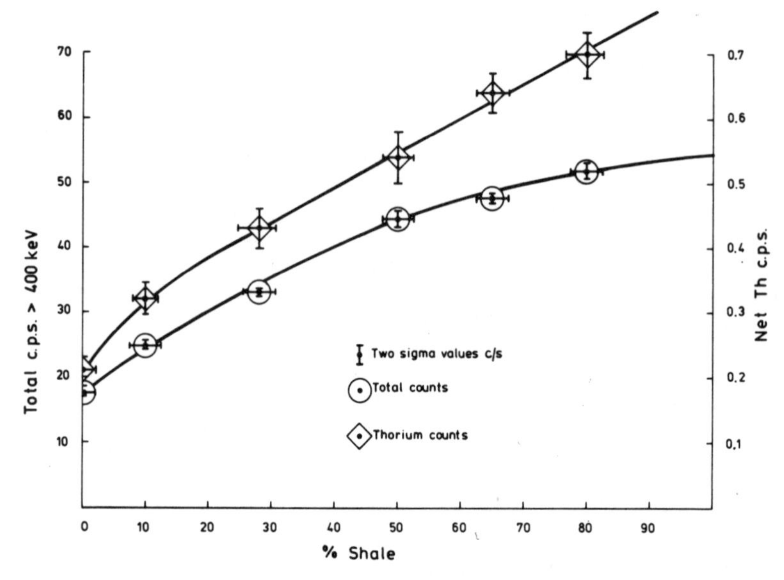

FIG.1. Natural gamma-ray response above 400 keV in synthetic mixtures of hematite and shale.

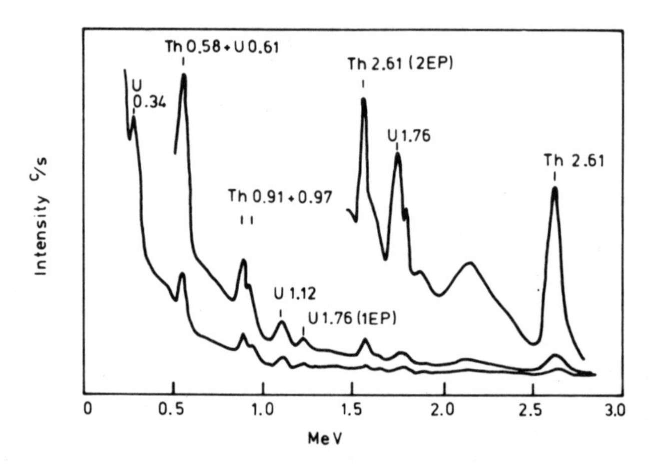

FIG.2. Typical borehole spectrum of mine-area shales. (Exploration hole E2; depth 12 m; 7.5 keV/channel.)

teletype (depth, count-rate, standard deviation of the count-rate) for control of the logging operations, and on punched tape for later regression analysis against assay values when the latter became available.

All the equipment was mounted in an insulated, air-conditioned caraven to counteract outside ambient temperatures that varied over the working period from 25 to 44°C and over a daily excursion of approximately half this range. The caravan temperature was maintained as close as possible to borehole temperature (23 to 25°C) to minimize waiting time for satisfactory functioning of the gain stabilizer after probe handling and calibration.[3]

Two bowspring centralizers were used on the logging probe. All holes were first calipered with a "clip-on" radiometric caliper device [8] or a commercial three-arm electromechanical caliper. Exploration holes averaged 14 cm in diameter with enlargements at the position of some shale bands or porous ore up to 18 cm. Blast holes averaged 32 cm with similar enlargements in diameter up to 35.5 cm. These changes in bore are not rapid and are not believed to cause any significant errors in measurement of natural radioactivity under prevailing conditions [4, 9] (see also footnote 1).

Results of this survey are summarized in Table I in the form of linear regression data.

3.3. Mine area sampling

In addition to the borehole logging investigations, further field sampling of shale bands exposed on the faces of various mining benches was carried out by collection of some forty 10-kg samples which were returned to the laboratory for crushing and counting of radioactivity. Samples were collected from every area of the mine and from every available shale band, as well as from some hematites adjacent to the shale bands. Visually identifiable sub-bands in thicker shale occurrences were also sampled (bands 9, 11, 14) using position and colour differences, as were a number of hematite "stringers" within such bands.

The 4-litre laboratory sample-container was of polypropylene plastic completely surrounding a high resolution 7.6 cm × 7.6 NaI(Tl) scintillator to give fixed and easily reproducible, cylindrical, geometry. All counting was carried out in a lead-shielded castle with a gain-stabilized multi-channel pulse height analyser system calibrated at 7.5 keV per channel. Counting proceeded in the same spectral regions as before by using essentially the same programme to obtain the net count-rates and corresponding standard deviations. Results of this survey are illustrated in Fig. 3.

4. DISCUSSION AND RESULTS

4.1. Laboratory results

Figure 1 presents the results obtained by measuring count-rate responses of various mixtures of hematite and shale in the spectral regions 400 keV - 3.0 MeV ("total response") and 2.44 - 2.71 MeV (thorium response). As predicted, there is an increasingly significant departure from linearity in the total response curve as the shale concentration increases. However, this departure is relatively small in the region-of-interest 0 - 50% shale (70 - 35% Fe). In field work, as already inferred, variations in specific activity of different shale bands preclude detection of any such departure from linearity of response. For practical purposes then, response may be taken as effectively linear over the range of interest.

---

[3] The range of the stabilizer was limited and too much warming of the probe outside the borehole or the caravan caused delays while control was re-established.

TABLE I. REGRESSION DATA FOR BLAST HOLES (B) AND EXPLORATION HOLES (E)[a]

Hole[b]	B5	B6	B7	B9	B10	All blast holes	E2	E3	All holes B and E
a_0	68.612	70.430		76.488	67.631	69.794	70.343	62.245	68.564
a_1	- 0.443	- 0.473		- 0.496	- 0.396	- 0.405	- 0.699	- 0.338	- 0.396
r	0.932	0.968		0.912	0.866	0.886	0.967	0.774	0.858
F	303.913	661.049		218.848	125.580	749.139	340.758	109.052	857.572
Mean counts/s	12.018	18.763	5.876	31.504	37.455	22.634	11.622	18.010	20.578
Mean % Fe	63.283	61.549	68.014	60.870	52.814	60.631	62.215	59.156	60.405
95% C.L. for prediction of mean grade, % Fe	± 0.515	± 0.611		± 0.809	± 2.649	± 0.688	± 1.013	± 1.007	± 0.574
95% C.L. for a single prediction at the mean grade, % Fe	± 3.163	± 4.232		± 5.548	± 17.767	± 9.928	± 5.264	± 8.778	± 10.087

[a] Data for hole B7 utilized only in pooled regression data, see text.

[b] a_0, a_1 coefficients of $\hat{y} = a_0 + a_1x$.

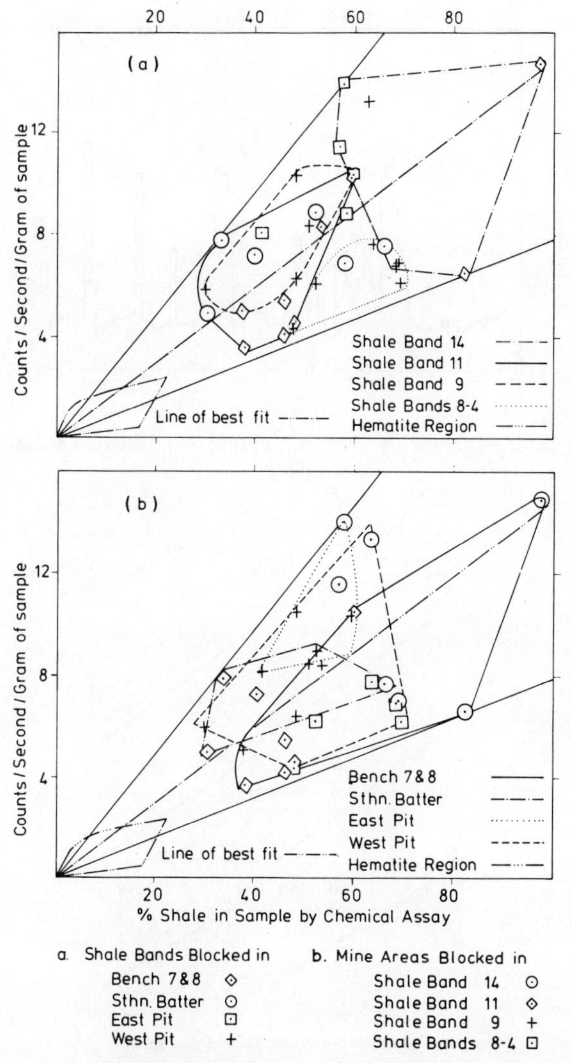

FIG. 3. Natural radioactivity of mine-area ferruginous shale samples.

The same conclusions apply to thorium counts (Fig. 1) for which it is difficult to assess whether curvature in the laboratory response curve is significantly less than for total counts owing to limited accuracy of the experimental data.

4.2. Field results, logging

Figure 4 shows a typical log of an exploration drill hole traversing horizontal bands of iron ore and shale. Blast holes are shallower (~ 18 m) and therefore result in exposure of only a few of the shale bands shown in Fig. 4 for the deeper exploration hole.

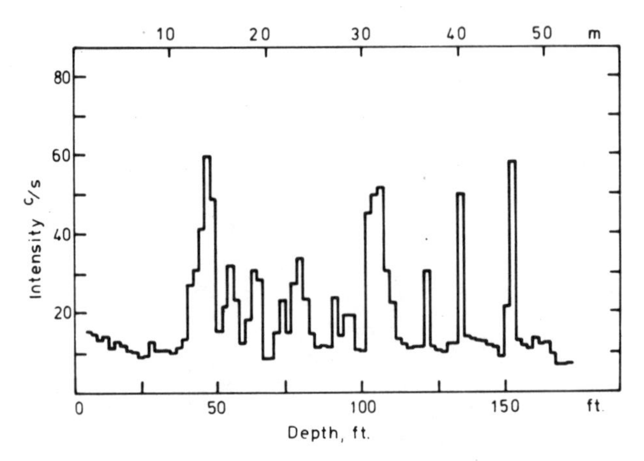

FIG.4. Log of natural gamma-ray activity in an exploration borehole. (Threshold 400 keV.)

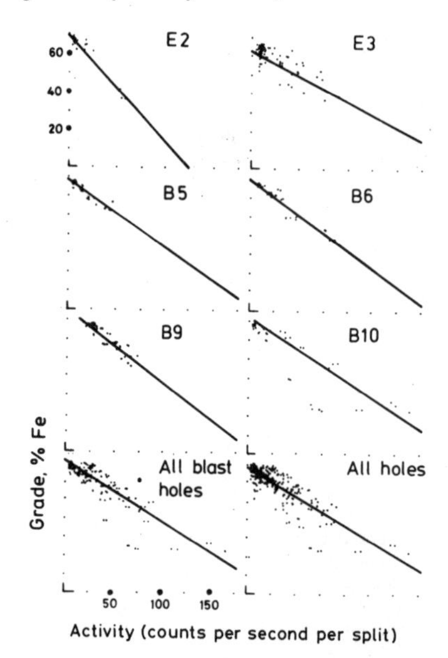

FIG.5. Linear regression analyses of various blast holes (B) and exploration holes (E) including pooled data (B) and (B+E).

The typical borehole spectrum of shale from the mine area (Fig. 2) may now be discussed. Although there is some variation in the ratio of uranium activity to thorium activity, in over 50 spectra of shale samples from widely differing localities taken either in boreholes or in the laboratory no significant above-background indications of potassium were found. Potassium concentration, therefore, is below the limits of detection by simple spectrometric techniques in an iron-rich shale matrix, i.e. its concentration is probably always less than 0.1% in shales from the mineralized zones[4].

[4] Potassium is present in substantial quantities, however, in the McCrae shales forming the footwall of the deposit.

Chemical analysis of 20 of the 40 samples referred to in Section 3.3 confirmed that potassium was, in fact, always less than 0.05% in the mine area. This is in accord with geological information that clay minerals in this mining area are essentially kaolinitic in character.

Observed radioactivity is therefore due to uranium and thorium. Values for each logging split have been plotted against grade for various holes in Fig. 5, together with lines fitted for each hole by the simple expression:

$$y = a_0 + a_1 \cdot x \quad (6)$$

where x = counts per second per split, y = grade in % Fe, a_0, a_1 are coefficients of the regression, r the correlation coefficient, F the F statistic, and 95% C.L. the 95% confidence limits for prediction of grade from radiometric assay (Table I).

Also plotted in Fig. 5 (lower left), and summarized in Table I, are pooled data from all blast holes, and, finally, pooled data from all available blast and exploration holes taken together (Fig. 5, lower right). Data for hole B7 was included in the latter regressions although the regression line for this hole is not significant by itself owing to the complete absence of shale bands traversed by the drill hole. (This results in strong clustering of the data.)

The regression lines for West Pit blast holes vary in slope (a_1 -0.396 to -0.496) with a_0 in the vicinity of its theoretical value of 70.0. The 95% C.L. for prediction of mean grade[5] from pooled radiometric assay results for all blast holes is ± 0.69% Fe despite the scatter of results for hole B10. The data pertain to two logging runs per hole; if only a single log is done instead of two, 95% C.L. (mean) for B5, for example, is ± 0.75% Fe compared with ± 0.52% Fe, and for all blast holes ± 0.86% Fe compared with ± 0.69% Fe. As indicated, all these figures pertain to prediction of mean grade. The scatter of data for single point prediction is such that confidence limits for such predictions are so wide as to be virtually useless (Table I). For example, even at the level of the mean grade the 95% C.L. for prediction of grade from a single radiometric observation is ± 10% Fe for the "all holes" regression.

It is also evident from Table I that parameters for the exploration holes taken together are consistent with the pooled data for blast holes, although, taken singly, holes E2 and E3 give values for a_1 which are "outliers" for the regression of pooled data for all holes.

The 95% C.L. (mean) for all holes logged is thus ± 0.57% Fe and the "pooled" expression serving as a calibration for predicting mean grade, y, from mean radiometric assay in an unknown hole is:

$$y = 68.56 - 0.3960\,x \quad (7)$$

with a correlation coefficient 0.86.

However, examination of residuals from the above plot indicates that the scatter of x values decreases as count-rate per split approaches zero, i.e. as grade approaches 70% Fe or 100% hematite. This suggests the desirability of a transformation of the data to obtain a sounder basis for evaluation of the confidence limits.

[5] Hereafter referred to only as "95% C.L. (mean).

Suitable transformations are:

$$z = \text{logit}\ (\%\text{Fe}) = \log_e\{y/(70-y)\},\ \text{and}\ y = \log_e x$$

where y is wt% Fe in the ore and x is the number of counts per second per split. The regression model fitted on this basis is:

$$z = \beta_0 + \beta_1 w + \beta_2 w^2 + \epsilon \tag{8}$$

where β_0, β_1, β_2 are the parameters of the model and ϵ is the increment by which any individual value of z departs from the regression curve. The predicted value of z for a given value of w is therefore:

$$\hat{z} = b_0 + b_1 w + b_2 w^2 \tag{9}$$

where b_0, b_1, b_2 are the least squares estimated values of β_0, β_1 and β_2. The estimated grade is given by:

$$\hat{y} = 70\ [1 + \exp(-b_0)\ x^{-(b_1 + b_2 \log_e x)}]^{-1} \tag{10}$$

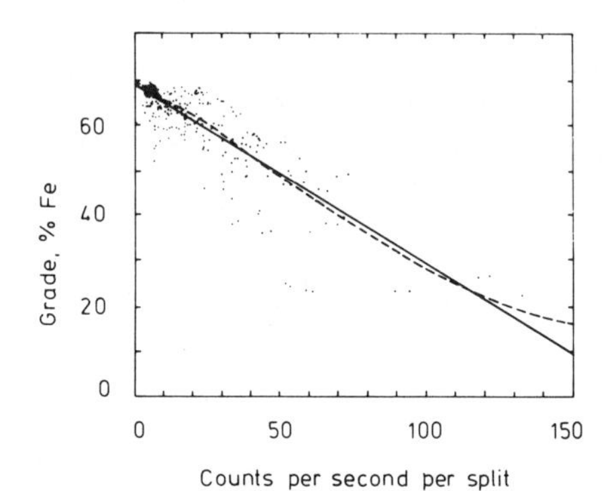

FIG.6. Comparison of two statistical models.
—— Simple linear model, Eq.(6)
---- Back-transformed model, Eq.(10).

An examination of this curve (Fig. 6) shows that it is virtually coincident with the straight line $\hat{y} = a_0 + a_1 x$ except in the relatively unimportant grade-range below 25% Fe[6]. For all practical purposes, therefore, the simple linear relationship (6) is both suitable and convenient for predicting iron grade from observed radiometric data.

Confidence limits derived from the simple linear relationship are also sufficiently precise for calculation of predicted grade values in the vicinity of the mean grade for the borehole to warrant their use in the interests of simplified computation (compare Tables I and II). If, however, confidence limits are required at lower (or higher) grades than the mean grade, the above transformations should be used as the basis of calculations. Details

[6] Few data points lie below this range.

TABLE II. REGRESSION DATA DERIVED FROM TRANSFORMATION

Type of regression [a]	All blast holes transformed	All holes transformed
b_0	- 2.842	- 2.911
b_1	- 0.797	- 0.515
b_2	0.329	0.266
r	0.888	0.831
F	381.992	341.453
Mean counts/s	22.634	20.578
Mean % Fe	60.631	60.405
95% C.L. for prediction of mean grade, % Fe	+ 0.806 - 0.870	+ 0.775 - 0.832
95% C.L. for a single prediction at the mean grade, % Fe	+ 5.812 - 12.222	+ 6.390 - 14.528

[a] b_0, b_1 and b_2 refer to Eq.(9).

are given in the Appendix. Confidence limits for several values in the range of counts per second per split derived from the simple linear form and from the transformations are compared in Tables III and IV.

4.2.1. Scatter of field data

The total scatter about the regression line derived in the previous section arises from numerous causes, only a few of which can be quantified with any degree of certainty. These can be conveniently divided into two broad categories:

(a) Physical and geophysical factors arising from the characteristics of the measuring equipment or the nature of the logging technique. Also included in this category are other factors such as variations in count-rate superimposed on normal counting statistics by variations in the specific activity of shale (q.v.).

(b) Chemical and geological factors arising from inadequacies in the assay data, or the natural variability of the materials concerned.

(a) Physical and geophysical factors

(a_1) Purely physical effects comprise random statistical errors due to the relatively short counting periods available and instrumental errors due to changes in gain of the electronic system, changes due mostly to zero shifts and slow, but largely unavoidable temperature changes in the probe.

Standard deviation of the count-rates for 61 cm (2 ft) splits at 45-60 cm/s logging speed (1.5 to 2 ft/min) varies over a range of 2 to 5%, being least in shale bands and greatest in hematite zones. As a fraction of the actual count-rate, however, this variation is a minor proportion of the total recorded. (It can, in any case, be reduced by the use of larger counting crystals.)

TABLE III. CONFIDENCE LIMITS FOR ALL BLAST HOLES
Comparison of values at different grades for two statistical models

Grade	% Fe = $a_0 + a_1x$		$z^a(\%Fe) = b_0 + b_1 w + b_2 w^2$ ($w = \log_e x$)	
	95% Limits for line	95% Limits for point	95% Limits for line	95% Limits for point
35	± 1.20	± 10.1	± 3.66	± 17.2
40	± 1.66	± 10.0	+ 3.02 - 3.09	+ 15.7 - 18.0
45	± 1.33	± 10.0	+ 2.36 - 2.46	+ 13.7 - 18.2
50	± 1.04	± 10.0	+ 1.75 - 1.84	+ 11.5 - 17.6
55	± 0.80	± 9.9	+ 1.23 - 1.31	+ 9.0 - 15.8
60	± 0.69	± 9.9	+ 0.85 - 0.91	+ 6.2 - 12.7
60.63	± 0.69	± 9.9	+ 0.81 - 0.87	+ 5.8 - 12.2
61	± 0.69	± 9.9	+ 0.78 - 0.84	+ 5.6 - 11.9
62	± 0.70	± 9.9	+ 0.71 - 0.77	+ 5.0 - 11.0
63	± 0.71	± 9.9	+ 0.64 - 0.70	+ 4.4 - 10.0
64	± 0.73	± 9.9	+ 0.56 - 0.61	+ 3.8 - 8.9
65	± 0.76	± 9.9	+ 0.47 - 0.52	+ 3.1 - 7.7
66	± 0.79	± 9.9	+ 0.37 - 0.40	+ 2.6 - 6.4
67	± 0.83	± 9.9	+ 0.26 - 0.28	+ 1.9 - 5.0

[a] See Section (b) of the Appendix.

TABLE IV. CONFIDENCE LIMITS FOR ALL BLAST AND EXPLORATION HOLES

Comparison of values at different grades for two statistical models

Grade	$\%Fe = a_0 + a_1x$		$z^a(\%Fe) = b_0 + b_1w + b_2w^2$ ($w = \log_e x$)	
	95% Limits for line	95% Limits for point	95% Limits for line	95% Limits for point
35	± 1.83	± 10.2	± 3.87	± 19.0
40	± 1.51	± 10.2	+ 3.12 - 3.22	+ 17.2 - 20.0
45	± 1.20	± 10.1	+ 2.38 - 2.48	+ 15.0 - 20.4
50	± 0.91	± 10.1	+ 1.70 - 1.79	+ 12.5 - 19.9
55	± 0.68	± 10.1	+ 1.16 - 1.23	+ 9.7 - 18.2
60	± 0.57	± 10.1	+ 0.80 - 0.96	+ 6.6 - 14.9
60.40	± 0.57	± 10.1	+ 0.78 - 0.83	+ 6.4 - 14.5
61	± 0.58	± 10.1	+ 0.74 - 0.79	+ 6.0 - 14.0
62	± 0.58	± 10.1	+ 0.67 - 0.73	+ 5.4 - 13.0
63	± 0.60	± 10.1	+ 0.60 - 0.65	+ 4.7 - 11.8
64	± 0.62	± 10.1	+ 0.52 - 0.56	+ 4.1 - 10.6
65	± 0.65	± 10.1	+ 0.42 - 0.46	+ 3.4 - 9.2
66	± 0.69	± 10.1	+ 0.32 - 0.34	+ 2.7 - 7.7
67	± 0.73	± 10.1	+ 0.35 - 0.40	+ 2.1 - 6.1

[a] See Section (b) of the Appendix.

Electronic drifts, which can also be reduced by improvements to the overall system, are difficult to quantify, but are, likewise, believed to make a minor contribution to the total variation in observed count-rate.

(a_2) Geophysical factors include the effect of non-coincidence between splits sampled by the logging probe and those recovered for chemical assay. A portion of the error due to this effect arises from the difficulty of operating in rough terrain from the same zero as that adopted by the drilling crew. Depth measuring errors can also arise in either drilling or logging operations, those due to logging being more easily identifiable and corrected in the regression analysis.

The combined effects, amounting to the offset of part or all of a complete split, result in a lowered probe response. The same effect can arise when strata dip at substantial angles to the axis of the borehole, and in both cases reproducibility and quality of fit of the regression line are degraded.

Regressions for quality of fit were therefore carried out both for probe response against nominally equivalent "assay split" depths as well as for depths offset by a single split. The best fit could then be selected without violating the principle of statistical independence of successive samples.

(a_3) Variation in specific activity of various shales is considered to be a potential source of scatter of radiometric assay results. The data points for chemical and radiometric assays of shale samples collected in various sections of the mine (cf. Section 3.3) are coded in Fig. 3 by shale band numbers and location. The distribution of these points, which were obtained under laboratory control and which are therefore essentially free from "logging" errors due to the factors enumerated above, clearly confirms that there are substantial variations in the specific activity of various shales.

Figure 3(a) shows boundaries of zones characterising distribution of points by band number[7]. If the single, outlying point for shale band 14 at 97% shale (3% hematite) is ignored as being a rare occurrence from a near-surface zone of the mine, it can be seen that it would be difficult to clearly identify the location of any new point by band number.

Figure 3(b) shows boundaries of zones characterising distribution of points by locations in the mine area. These areas are geographically well defined and clearly separated from each other. If the single outlying point at 97% shale is again ignored, it can be seen that it would be difficult to clearly identify the position of any new data point by zone location.

The mean grade for all points is calculated to be 42% shale ($\sim$ 6 counts per second per gram), which makes it fruitless to calculate the 95% C.L. for mean grade-of-interest of a borehole-type sample, as this is located at a mean of $\sim$10% shale (90% hematite or $\sim$ 63% Fe) and cannot be calculated from the above data with sufficient precision.

However, the data obtained by logging in the West Pit area encompass all major shale band occurrences (bands 9, 11 and 14) while Fig. 3 shows that the spread of data from the West Pit is virtually as great as that observed in any section of the mine for which test material was obtainable. Data from new holes will be derived from the same set of shale bands in basically the same or adjacent areas, and would therefore be expected to yield count-rate data producing mean grades consistent with those cited in Tables I and II.

[7] In the mine, 21 shale bands are numbered in sequence commencing with band 4 at the base of the mineralised zone [13].

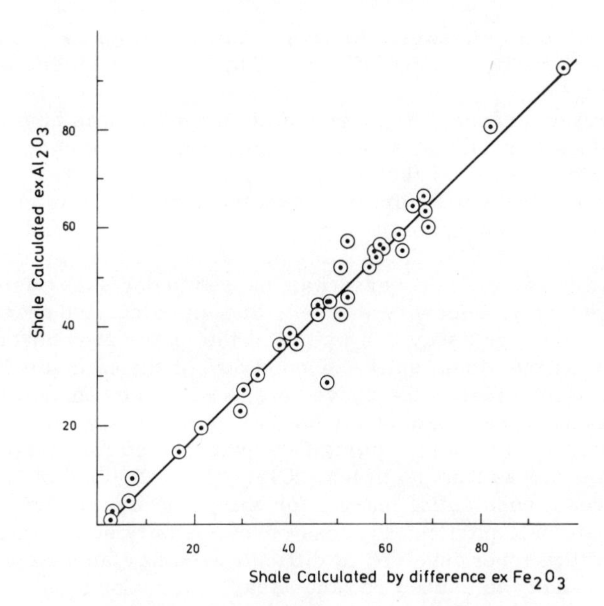

FIG.7. Chemical assays for shale (kaolinite) taken from mine area.

(a_4) Extraneous material such as quartz in a system assumed to consist essentially of hematite and shale occupies space in the ore which in the present context has been assumed to be occupied by shale having a given mean specific radioactivity. Sporadic occurrences of this material, therefore, will also increase variability in the count-rate. This variability, however, is now shown to be sensibly constant throughout the mine area, although it may occasionally show a larger variation for a given split or even series of several splits, particularly in drill holes approaching the unmineralised banded-iron formations.

Figure 7 shows a plot of alumina, y, determined by chemical assay, and calculated to shale (kaolinite) for 40 samples of shale referred to in Section 3.3 against kaolinite, x, calculated as the difference: (100-hematite) where hematite has also been determined accurately by chemical assay for iron. The regression expression for this data is:

$$y = 0.959x - 1.65$$

with r = 0.99, F = 2004. Thus, the difference between actual shale content and that attributed by difference is not constant (- 1.65%), but increases somewhat with increasing shale content. In terms of iron grade this average "mine area" difference amounts to 2.15% at 63% Fe and 2.56% at 56% although, as indicated by the figure, the difference may be greater in individual samples (splits).

(b) Chemical and geological factors

(b_1) Assay errors. A repeated routine test assay of material recovered from a blast hole gave a mean differing insignificantly from the mean of the first set of assays. Assays of individual splits, however, were observed to vary within a range of ± 1%. These errors enter into the regression.

(b_2) Sampling errors. Recovery of diamond cores is high in solid ore (> 95%) but less so in friable, shaly or highly goethetised sections, where recovery falls to 50 - 70% of the core.

In percussion and rotary drilling, however, recovery is never complete for any split due to poor uphole trapping of some particles and loss of lighter particles in dust created by the drilling operation. Thus, in one exploration hole where every sample recovered was bagged under supervision and weighed before assay, recovery was only 36% as calculated from the average hole diameter, average assay and weight of the material recovered. In fact, no recovery of an individual split exceeded 50% of the calculated value. Of the unrecovered material in the above hole, 40% was estimated to have accumulated around the mouth of the hole as an obviously heavy material rich in coarse hematite particles, leaving 24% unaccounted for and presumably dispersed in the air as fine particles. Clearly this method of "sample splitting" leaves a substantial margin for sampling errors due to preferential loss of fine material, particularly loss of low-density shale particles. Assessment of the error involved is difficult without elaborate experimentation but seems likely to exceed routine assay errors for most splits and show a bias towards high iron values.

For blast holes, where present mine practice is confined to collection of a single shovel-full of material weighing less than 2 kg from a grossly segregated heap of cuttings weighing more than 4000 to 5000 kg, the 95% confidence limits of the sampling error for mean grade are believed to be of the order of ± 5% Fe.

Another potential source of logging errors, the smearing or entrapment of material from one split over the surface of the borehole above that split is not considered to be a sampling problem of significance in this work as excessive moisture is uncommon on the area and holes can be drilled without fluid drilling-aids and can be properly "blown out" with compressed air to shift accumulations of fine material from the walls.

(b_3) Geological variability. Of importance in all logging operations is the geological variability of the material whose assay (grade) is being sought, for unless variability is insignificantly small the identity in composition of material around the borehole and within the borehole cannot be assumed. Quantitative expression of this situation has been given by Matheron [10] in his use of the concept of "extension geovariance" to describe errors arising from extrapolation of assay data from one drill hole to others in an exploration grid [11]. The same errors arise on the smaller scale involved in calibration of logging probes and interpretation of logging data. For instance, it has been shown by Aylmer and Wylie [12] that rings of diamond core holes drilled in large blocks of iron ore (4 to 6 tonnes) around a central 5-in (12.7-cm) diamond or percussion drill hole (and into the zone of influence of a logging probe) lead to recovery of material whose composition can differ increasingly from adjacent core material as the grade of hematite diminishes. Similarly, and more relevantly for a logging situation, an annulus of material drilled around the central percussion hole, or a second 12.7-cm drill hole only 18 cm away from the first hole, can lead to recovery of material differing in composition from that in the first drill hole by at least 1 - 2% Fe for material of average grade 63% Fe. These differences are smaller in high grade ore and increase as variability increases or grade diminishes. They are expressions of grade variability (i.e. in effect of shale distribution) which are independent of variable distribution of radioactivity in the shales.

5. SUMMARY

The natural radioactivity of uranium and thorium present in layered mixtures of iron ore and shale has been utilized to determine the grade of ore.

The scatter of field data points (the greatest single cause of which is believed to be a specific activity of shale varying from place to place and from one shale band to another about a mean value) conceals any predicted, small, departure from linearity of detector response which, in practice, is adequately approximated by a simple linear expression connecting ore grade with observed radioactivity. This expression permits prediction of the mean grade of a group of blast and exploration boreholes with 95% confidence limits based on the linear model of ± 0.6% Fe at the 63% Fe level. Confidence limits based on single split observations are too wide to be of practical use.

More accurate values of the confidence limits may be obtained by transformation of the data, but the additional computation involved is considered to be unwarranted for routine application unless confidence limits at grades differing substantially from that of the mean are desired.

The method, which requires no external radioactive source, is simpler to apply and less stringent in its demand on technical resources than alternative logging methods producing approximately equivalent confidence limits for grade prediction. Data output can be immediately processed for incorporation into mine production and planning calculations.

ACKNOWLEDGEMENTS

Particular thanks are due to the staff of Hamersley Iron Pty. Ltd. at Mt. Tom Price for their unfailing co-operation in the prosecution of this work. The many helpful discussions with Dr. R. Jarrett and Mr. R. Forrester, CSIRO, Division of Mathematics and Statistics, are gratefully acknowledged as is the valuable assistance of Mr. S. Youl in the processing of field data.

REFERENCES

[1] McLEOD, W.N., The geology of the iron deposits of the Hamersley Range Area, Western Australia, Western Australia Geol. Surv., Bull. 117 (1966).

[2] TRENDALL, A.G., BLOCKLEY, J.G., The iron formations of the Precambrian Hamersley Group, Western Australia, Western Australia Geol. Surv. Bull. 119 (1970).

[3] JONES, M., WALRAVEN, F., KNOTT, G.G., Natural gamma logging as an aid to iron ore exploration in the Pilbara region of Western Australia, Aust. Inst. Min. Metall., W.A. Conference Proc. (May, 1973) 53.

[4] CZUBEK, J.A., Some problems in the theory and quantitative interpretation of natural gamma ray logs, Acta Geophys. Pol. 9 (1962) 121.

[5] GOLDSTEIN, H., WILKINS, J.E., Calculation of the penetration of gamma rays, USAEC Rep. NYO-3075 (1954).

[6] EISLER, P.L., HUPPERT, P., A nuclear geophysical borehole logging system, to be published.

[7] EISLER, P.L., LEWIN, F., "On-line computer control of borehole logging", IFAC Symposium on Automatic Control in Mining and Metal Processing, Sydney, 1970. Institute of Engineers, Australia, National Conference Publication No.73/4 (i), p.87.

[8] CHARBUCINSKI, J., WYLIE, A.W., Radiometric calipers for borehole logging, to be published.

[9] HYMAN, S.C., MINUSKIN, B., CERTAINE, J., How drillhole diameter affects gamma ray intensity, Nucleonics 13 (1955) 49.

[10] MATHERON, G., Principles of geostatistics, Econ. Geol. 58 (1963) 1246.

[11] BLAIS, R.A., CARLIER, P.A., Application of geostatistics in ore evaluation, Trans. Can. Inst. Min. Metall. Spec. 9 (1968) 41.

[12] AYLMER, J.A., WYLIE, A.W., An examination of grade, density and loss-on-ignition values distributed around percussion drillholes in hematite blocks from Mt. Tom Price, Western Australia, Investigation Report 503 R (1972), Division of Mineral Physics, CSIRO.

[13] CRONE, J.G., HAMMOND, J.R., WARD, T., Quality control at Hamersley Iron, Aust. Inst. Min. Metall., W.A. Conference Proc. (May, 1973) 313.

APPENDIX

In general, the 95% confidence limits of the predicted mean value of z, $\hat{z}_k$ at a specified value w_k of the independent variable, w, may be expressed as:

$$2_k \pm t\sqrt{Var(\hat{z}_k)} \tag{1}$$

where t is the t(ν, 0.975) statistic for the 95% confidence level, and ν is the number of degrees of freedom for the residual sum of squares, i.e. (n - 2) for the linear model and (n - 3) for the quadratic model.

For a single new observation, w_j, the variance of its predicted value $\hat{z}_j$ is:

$$Var_p(\hat{z}_j) = s^2 + Var(\hat{z}_j) \tag{2a}$$

where s^2 is a pooled, unbiased estimate of the variance σ^2, so that the 95% confidence limits in this case are:

$$\hat{z}_j \pm t\sqrt{(s^2 = Var\,\hat{z}_j)} \tag{2b}$$

For a set of q new observations, $w_1, \ldots, w_q$, as in the case of a new borehole with q splits,

$$\bar{z} = \left(\sum_{j=1}^{q} z_j\right) \Big/ q \tag{3a}$$

$$Var_p(\bar{z}) = \left\{\sum_{j=1}^{q} [s^2 = Var(\hat{z}_j)]\right\} \Big/ q^2$$

with the 95% C.L. $\bar{z} \pm t\sqrt{Var_p(\bar{z})}$ (3b)

The tolerances of prediction have been applied to two statistical models in this paper:

(a) The simple linear model

$$\hat{y}_i = a_0 + a_1 x_i$$

Thus $z_i = y_i$ (%Fe), and $w_i = x_i$ (counts per second per split), and $\mathrm{Var}(\hat{z}_j) = \mathrm{Var}(\hat{y}_j)$

$$= s^2 \left\{ 1/n + (x_j - \bar{x})^2 \Big/ \sum_{i=1}^{n} (x_i - \bar{x})^2 \right\}$$

The 95% confidence limits are:

$$\hat{y}_j \pm t\, s \sqrt{1/n + (x_j - \bar{x})^2 \Big/ \sum_{i=1}^{n} (x_i - \bar{x})^2}$$

For predicting the mean grade of a new hole with q splits, the 95% confidence limits for the mean grade are:

$$\bar{y} \pm t\, s \sqrt{\left\{ \sum_{j=1}^{q} [1 + 1/n + (x_j - \bar{x})^2] \Big/ \sum_{i=1}^{n} (x_i - \bar{x})^2 \right\} \Big/ q^2}$$

(b) The transformed quadratic model

$$z_i = \log_e [y_i / (70 - y_i)], \text{ and } w_i = \log_e x_i$$

where

$$z_i = b_0 + b_1 w_i + b_2 w_i^2$$

This is more conveniently rewritten as:

$$\hat{z}_j = c_0 + c_1 (w_j - \bar{w}) + c_2 \{(w_j - \bar{w})^2 - S_2/n\}$$

so that for the mean predicted value $\hat{z}_j$ for a specified w_j, the variance is

$$\mathrm{Var}(\hat{z}_j) = s^2 \,[1/n + v_1 (w_j - \bar{w})^2 + 2v_2 \{(w_j - \bar{w})^3 - S_2 (w_j - \bar{w})/n\}$$
$$+ v_3 \{(w_j - \bar{w})^4 - 2S_2 (w_j - \bar{w})^2/n + S_2^2/n^2\}]$$

and the confidence limits are as given by Eq. (1).

For the mean of q new observations, as for the q splits of a new hole:

$$\mathrm{Var}_p(\bar{z}) = (s^2/q^2)\left\{\sum_{j=1}^{q}[1 + \mathrm{Var}(\hat{z}_j)]\right\}$$

as in Eq. (3a), where

$$S_k = \sum_{i=1}^{n}(w_i - \bar{w})^k$$

and regression is based on n observations.

$$D = S_2 S_4 - S_2^3/n - S_3^2$$

$$v_1 = (S_4 - S_2^2/n)/D$$

$$v_2 = -S_3/D$$

$$v_3 = S_2/D$$

The confidence limits are as expressed in Eq. (2b) for a single split (i. e., for $q = 1$) and as in Eq. (3b) for numerous splits.

So as to usefully transform back to grade, we refer to each pair of confidence limits in the transformed frame of reference as h_i ($i = 1, 2$).

Thus, the corresponding values of grade are:

$$y_i = 70 \exp(h_i)/[1 + \exp(h_i)]$$

IN-SITU CAPTURE GAMMA-RAY ANALYSES FOR SEA-BED EXPLORATION

A feasibility study

F.E. SENFTLE, A.B. TANNER, P.W. PHILBIN
US Geological Survey,
Reston, Virginia

J.E. NOAKES, J.D. SPAULDING, J.L. HARDING
University of Georgia,
Athens, Georgia,
United States of America

Presented by J.R. Rhodes.

Abstract

IN-SITU CAPTURE GAMMA-RAY ANALYSES FORSEA-BED EXPLORATION: A FEASIBILITY STUDY.

In-situ neutron-capture gamma-ray spectra of bottom sediments were taken at five stations in western Long Island Sound using a californium-252 neutron source and a Ge(Li) detector. Chlorine, in addition to being a major source of interference, did not yield the same spectrum when measured in sea-water and in the bottom sediments. Other anomalous results were observed when the spectra of the bottom sediments from the various stations were compared. After analysis of these results in the light of the spectral distribution of neutrons obtained from Monte Carlo calculations, it appears that the measured capture gamma-ray spectra are the result of both thermal and epithermal resonance capture. The results of laboratory experiments tend to confirm this hypothesis. Substantial epithermal capture poses a serious problem of identification and calibration for quantitative analysis. The use of many small sources arranged around a central detector should improve the counting statistics and reduce the seriousness of the problem.

INTRODUCTION

Throughout the world, rapidly growing industrialization is causing an ever-increasing comsumption of raw materials. For this reason, the seabed is gradually drawing increased attention as a source of some of the elements that are in short supply. In addition to the impending mineral crisis, industrial centers are dumping large quantities of potential raw materials--and pollutants--into the rivers and oceans. We must increase our knowledge of the distribution of the elements on the ocean floor and be able to monitor the changes in concentration with time. To do this, an in situ technique would be most useful. Three nuclear methods have been proposed as possible solutions to the problem: delayed gamma-ray spectroscopy similar to that used by Noakes and Harding[1], capture gamma-ray spectroscopy, as tested by Moxham et al[2]; and X-ray fluorescence spectroscopy, as suggested by Cooper et al[3].

In July 1974, several agencies of the U. S. Government[1], the Battelle Pacific Northwest Laboratories, and the University of Georgia, pooled their efforts to field test and compare delayed

[1] Atomic Energy Commission, National Oceanic and Atmospheric Administration, US Geological Survey, and National Aeronautics and Space Agency.

TABLE I. GEODETIC STATION POSITIONS AT WHICH MEASUREMENTS WERE MADE[a]

Station	Date	Position
1A	23 July	40° 53' 19" 73° 44' 51"
1B	24 July	40° 53' 17" 73° 44' 48"
2	25 July	40° 53' 07" 73° 44' 37"
3	26 July	40° 42' 44" 73° 44' 45"
4	26 July	40° 52' 45" 73° 44' 50"

[a] Determinations by Lt. Commander Floyd Childress, NOAA.

TABLE II. PARTIAL CHEMICAL ANALYSES OF THE BOTTOM SEDIMENTS AT THE FOUR STATIONS

Station	1A	2	3	4
		Major Analyses (percent)		
Silicon	23.34	23.51	23.40	24.93
Aluminum	6.26	6.18	6.31	6.10
Iron	3.99	3.95	3.99	4.37
	4.37	4.09	4.09	4.48
Magnesium	1.41	1.37	1.39	1.24
	1.42	1.38	1.47	1.24
Calcium	1.25	1.14	1.47	1.57
	1.11	1.21	1.32	1.61
Potessium	1.95	1.91	2.06	1.83
	2.03	1.95	1.95	1.73
Phosphorus	0.08	0.06	0.06	0.06
Manganese	0.08	0.08	0.07	0.08
	0.08	0.08	0.07	0.07
Titanium	0.34	0.34	0.32	0.37
Sulfur	1.07	1.21	1.01	1.03
Carbon	4.05	3.31	2.91	2.78
		Minor Analyses (ppm)		
Copper	207	205	209	190
	207	200	202	190
Zinc	353	363	365	353
	344	350	365	359
Strontium	222	195	222	195
	195	167	167	138
Cadmium	104	88	84	96
		Trace Analyses (ppb)		
Gold	<100 ppb			

and capture gamma-ray spectroscopy and X-ray fluorescence methods of *in situ* elemental analysis of marine sediments. These tests were conducted in western Long Island Sound, about 20 km northeast of downtown New York City, between David's Island and Execution Rocks. We report here a feasibility study on the capture gamma-ray method only. The possible use of neutron-capture gamma-ray spectroscopy for marine exploration was first suggested by Senftle et al[4] and Wiggins et al[5]. However, in their studies no *in situ* measurements were made. The laboratory studies, however, were encouraging and indicated that this technique could be developed into a practical tool. Moxham et al[2] have since reported on some preliminary *in situ* capture gamma-ray measurements in an estuarine enviroment. We outline here a more detailed study of the problems to determine whether *in situ* measurements are feasible and to make recommendations for future experiments.

SAMPLING AND SITE DESCRIPTION

Measurements were made at four different sites (Table I). The water depth ranged from 15 m to 30 m, and the bottom muds at each site were black anaerobic sediments about 1 metre thick. Core samples and grab samples were taken at each site for more detailed studies. The chemical analyses are shown in Table II.

INSTRUMENTATION

The experiments were conducted from a barge, which was moved from station to station. A 4096-channel analyzer, minicomputer, magnetic tape recorder, and other associated electronic instruments were housed in a truck secured to the deck. Apparatus consisting of a neutron-source probe attached to a sealed fiberglass drum containing a Ge(Li) detector was lowered by winch over the side and was constructed to penetrate the sediments (see Figure 1). Californium-252 sources of 0.7 and 8.2 micrograms were each inserted into one of several ports in the probe. As shown in Figure 2, the source may be placed in six possible positions. In these experiments, the 8.2-μg source was placed 62.2 cm and the 0.7-μg source, 52.1 cm below the detector. These positions were chosen on the basis of the neutron flux density at the detector. In the designated positions, the neutron flux density exceeding 0.3-eV energy was 4 to 8 neutrons per cm^2-sec at the detector position when the apparatus was in the water. The magnitude of the flux density was well below the level which might incur damage to the germanium diode. As will be shown below, the size of the ^{252}Cf sources was an unfortunate choice which resulted in poor statistical results. A stronger source would have given better counting statistics without resulting in undue damage to the detector.

The Ge(Li) detector was located in the large upper section of the apparatus. Immediately above the source section of the probe and in line with the detector was a 6.9-cm-diameter-by-16.8-cm-long tin shadow shield. Tin was selected for the shadow shield because of its low interference parameter as determined by previous measurements[6]. The 45-cm^3 Ge(Li) detector was placed about 1.5 cm above the top of the shadow shield. The diode was kept at low temperature by a canister-cooled cryostat[7]. Both solid chlorodifluoromethane and propane were used for cryogens. The canisters, which were 7.9 cm in diameter and 20.3 cm long, were prefrozen on deck with liquid nitrogen and inserted during operating periods to maintain proper cooling (Figure 3). The detector, cryostat, and preamplifier were housed in a 41.4-cm-diameter stainless-steel pressure vessel.

The raw data were processed by a minicomputer using a sliding digital filter program written by Tanner[8]. This program removes the constant and linearly varying components of

FIG.1. ^{252}Cf-Ge(Li) probe being lowered over the side of a barge into the water.

the spectrum and retains only those peaks that exceed twice the standard deviation, σ, due to the statistical uncertainties of counting and the error associated with the computation.

It was anticipated that chlorine in the seawater would be a source of interference; therefore, the pressure vessel containing the detector was surrounded with a 50-cm diameter fiberglass shell filled with fresh water. The 17.8-cm sheath of fresh water around the detector was used to reduce possible chlorine interference.

The apparatus had a slight positive bouyancy, and for most of the experiments, negative bouyancy was assured by use of lead ballast weights. While on deck, the probe fit into an annular shield filled with fresh water. The shield was constructed so that a section could be removed to load or unload the sources. This operation is shown in Figure 4.

Additional gamma-ray and neutron measurements were made in the laboratory in a 5.5-m-diameter swimming pool to check certain aspects of the problem. These are described below.

FIG.2. Close-up detached source end of the probe showing the ports into which ^{252}Cf sources can be inserted.

POSITION OF THE SAMPLE

It was planned to lower the apparatus to the bottom so that the probe penetrated the sediments and the fiberglass drum would be more or less flush with the bottom. In this mode it was desirable to know whether the sample "seen" by the detector is principally around the source or near the water-sediment interface. Some experiments were therefore made on deck with the apparatus mounted in its water shield. A 700-g box of NaCℓ was placed at various positions along the probe (Figure 5A). The counts in the 5.600- and 5.089-MeV chlorine peaks were accumulated at each position during a 17.1-minute irradiation. The results shown in Figure 5B indicate that the most of the measured gamma radiation would be from those sediments directly below the large detector section of the probe, i.e. the surficial sediment.

THE CHLORINE PROBLEM

If one includes the single- and double-escape peaks, more than 400 gamma-ray peaks are possible in the chlorine spectrum. Of course, not all have sufficient intensity to be a source of interference. About 50, however, have full-energy intensities greater than 0.5 per 100 gamma-rays emitted. To help reduce the interference from chlorine in the seabed spectra, the fresh-water

FIG.3. Inside view of pressure vessel showing a solid propane canister being loaded into the cryostat.

FIG.4. View of probe resting in deck shield. Shield section has been removed to allow loading of source into nose section.

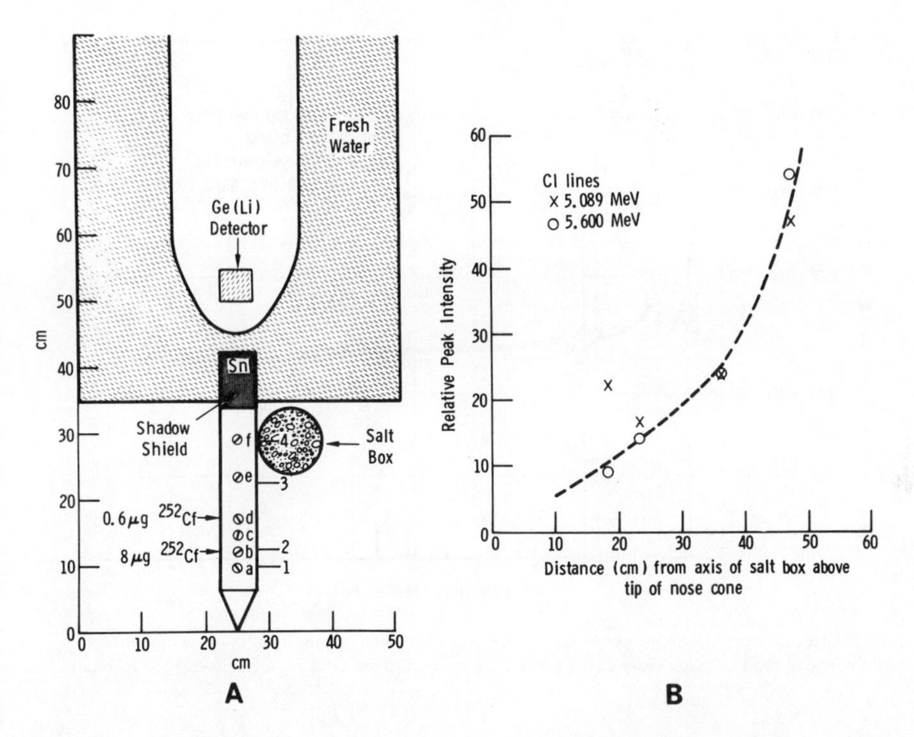

FIG.5. (A) Sketch of apparatus showing relative position of Ge(Li) detector, nose section and NaCl sample. (B) Relative peak intensity of the 5.089- and 5.600-MeV lines of chlorine as a function of distance above the deck (tip of nose section).

shield described above was used around the detector. A spectrum of the seawater was measured when the apparatus was suspended about 6 m below the surface. A total of 62 peaks were observed whose intensity was 2σ or more above the continuum. Of these, 48 were within 5 keV of a known chlorine peak and must be considered as chlorine peaks or peaks to which chlorine is a major contributor. Even with the high resolution of the Ge(Li) detector, the presence of so many chlorine peaks seriously limits the usefulness of the capture gamma-ray method of analysis. As resolution and stability of semiconductor detectors improves, this problem may become less acute. At this stage, we must be content to confine analyses to those energy windows in which there is little or no chlorine interference.

One obvious technique for the reduction of chlorine interference in a seabed spectrum is the channel-by-channel subtraction of a seawater spectrum taken using the same apparatus and making no changes in instrumental settings. This background subtraction method is inadequate in high-resolution capture gamma-ray analysis for at least two reasons: the great stability required for the instrument system is impractical, and the subtraction operation involves statistical errors so large that the result is complete loss of some real peaks and possible introduction of some false peaks.

Instrument stability requirements may be assessed approximately by considering that an energy range of from less than 0.5 to at least 9.0 MeV should be covered, or about 0.002 MeV per channel of a 4096-channel analyzer. At the high-energy end of the range,

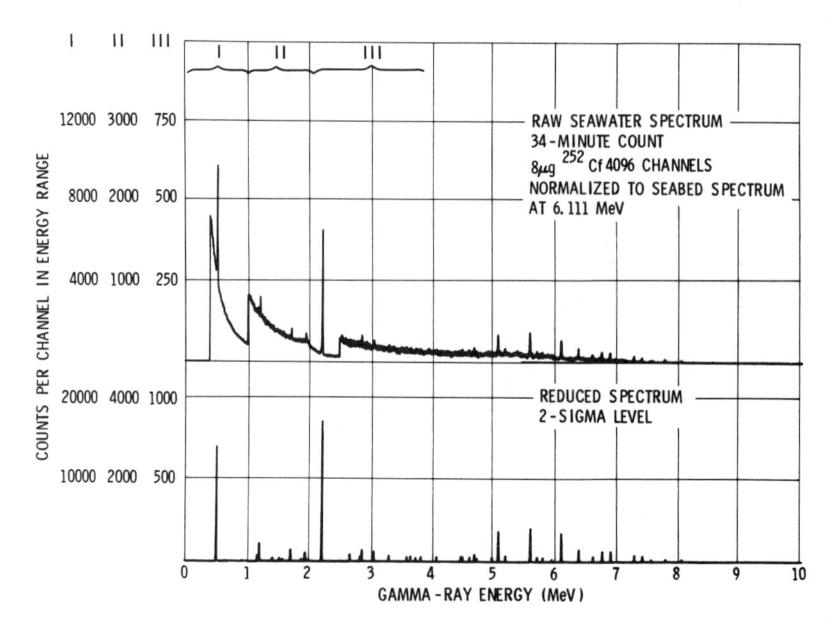

FIG.6. Untreated and reduced spectra of sea-water taken with probe suspended 20 m below the surface. Spectrum scaled so that the intensity of the 6.111 MeV is equal to the same peak in the sea-bed spectrum, Fig.7.

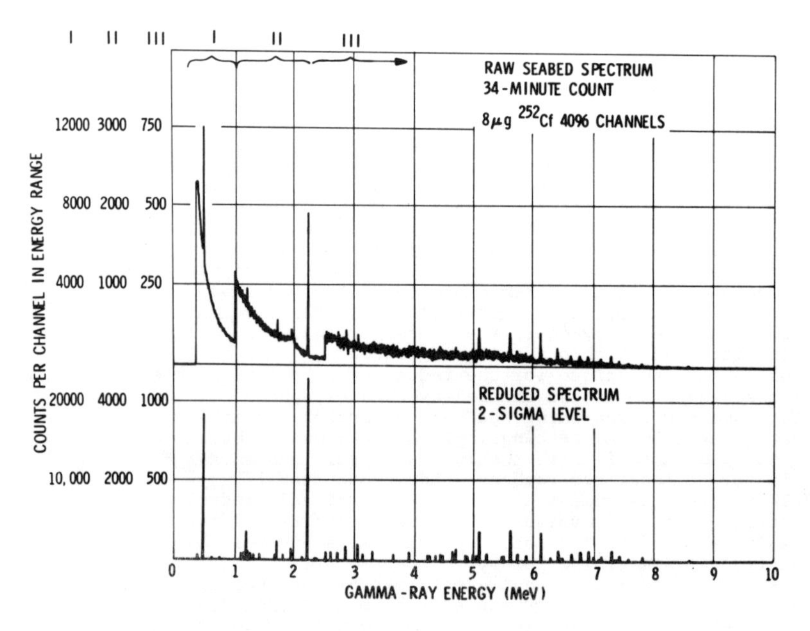

FIG.7. Typical untreated and reduced spectra of bottom sediments at Station 1A.

where a resolution of 0.010 MeV full width at half maximum can be realized, most of a peak is contained in five channels, and only one or two channels are near peak amplitude. If a peak centroid in the background spectrum differs from that in the "sample" spectrum by only half a channel, it represents an instability of only 0.012 percent full scale, yet it is sufficient to cause spurious peaks and valleys to appear in a spectrum processed by channel-by-channel background subtraction. To accomplish the background subtraction without introducing such distortions, a much more sophisticated technique must be used that adjusts the spectra to each other or that correlates peaks and finds the differences between their respective areas. The background spectrum must also be a high-quality one, relatively free of statistical error. Such a spectrum can be obtained by accumulation over a much greater period of time than that used for the "sample" spectrum and reduction by a constant fraction, but taking long background counts routinely or rigid standardization of the counting conditions may be quite impractical in the field.

THE PEAK INTENSITY

The untreated and reduced seawater spectrum and seabed spectrum taken at Station 1A are shown in Figures 6 and 7. The seawater spectrum was scaled to make the amplitude of the 6.111-MeV ^{35}Cl peak equal to that in the seabed spectrum. This was done to correct for difference in chlorine concentration and variation in the neutron flux density. Some of the chlorine peaks visible in the spectrum of seawater do not show in the spectrum of bottom sediments. This is easily seen by comparing the two reduced spectra. Examination of the number of counts in those peaks in the seawater spectrum that are missing in the seabed spectrum shows that in essentially every peak the computed statistical error is large enough to cause the peak to be not significant at the 2σ level; thus that peak does not appear in the reduced spectrum. Furthermore, the intensity ratios of peaks of the same energy in the normalized spectra sometimes deviate from unity by an amount greater than would be expected from statistical variation. The ratios of even some high-intensity peaks were outside the expected range. For instance, the intensity ratios of the 5.715- and 7.281-MeV peaks in the seabed to seawater spectra were 1.44 and 1.40, respectively. In addition to this problem, several peaks were at energies that did not correspond to any plausible radiative thermal-neutron-capture gamma rays. Although the statistical control in these in situ measurements was less than desirable for this kind of analysis, inconsistences from other causes appear to be superimposed on the poor statistics.

The seabed spectra taken at the other stations are dominated by chlorine just as is that shown in Figure 7 for Station 1A. Again the counting statistics were poor and only peaks with intensities above 2σ were considered. Although some peaks other than those of chlorine, such as Ni, Fe, Cu etc., could be qualitatively identified, quantitative analyses could not be made. Many of the peaks, even those with a rather large number of counts, did not appear in the spectra at all the stations. This was surprising, as the sediments differed little in composition from station to station.

COMPOSITIONAL EFFECTS ON NEUTRON FLUX DENSITY AND ENERGY DISTRIBUTION

As long as the sediments do not contain a significant concentration of some high cross-section element, the total neutron flux density can be expected to be higher in the sediments

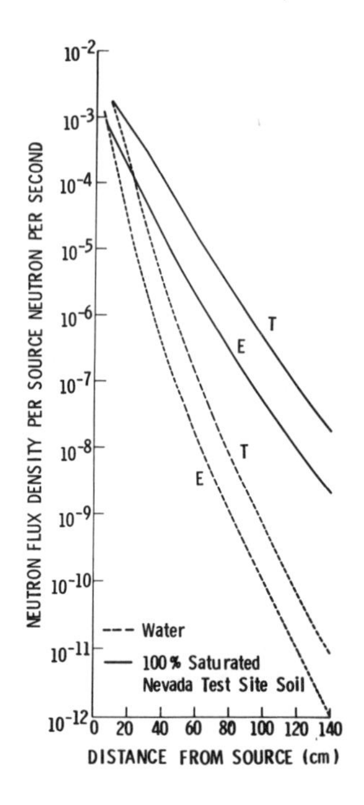

FIG.8. Monte Carlo calculation of the thermal (T) and epithermal (E) neutron flux density in water compared to that in 100% saturated soil of Nevada Test Site composition.

than in the seawater. Our measurements at all four stations are consistent with a significantly higher total neutron flux density in the sediments. This phenomenon is confirmed by the fresh-water Monte Carlo computations shown in Figure 8, i.e the total neutron flux density would be expected to be greater in the sediments than in the seawater. Similarly, experiments in a fresh-water pool in the laboratory showed the total neutron flux density to be greater within 15 cm of the underlying concrete floor than farther above it, where the environment was essentially an infinite water medium.

Figure 8 shows the results of a Monte Carlo computation, using the ANISN code, of slow (0-4 eV) and epithermal (4 eV to 15 keV)[2] neutron populations as a function of distance from a point source of ^{252}Cf in fresh water[9]. Beyond the point where equilibrium has been established (about 10 cm from the source), the ratio of thermal to epithermal neutrons is essentially constant, the thermal flux density being everywhere several times higher than the epithermal flux density. In saturated soil, the thermal- to epithermal-neutron flux-density ratio is slightly greater than for pure water.

Normalization of the spectra such as shown in Figures 6 and 7 corrects for differences in total neutron flux density but does not correct for any differences in neutron energy distribution in the sediments in comparison with the seawater. In water, or even in water-saturated sand, the epithermal-neutron flux density is generally about a tenth of the thermal-neutron flux density (see Figure 8). It is tacitly assumed that radiative neutron capture

[2] The energy ranges are not strictly thermal and epithermal but are sufficiently so for this discussion.

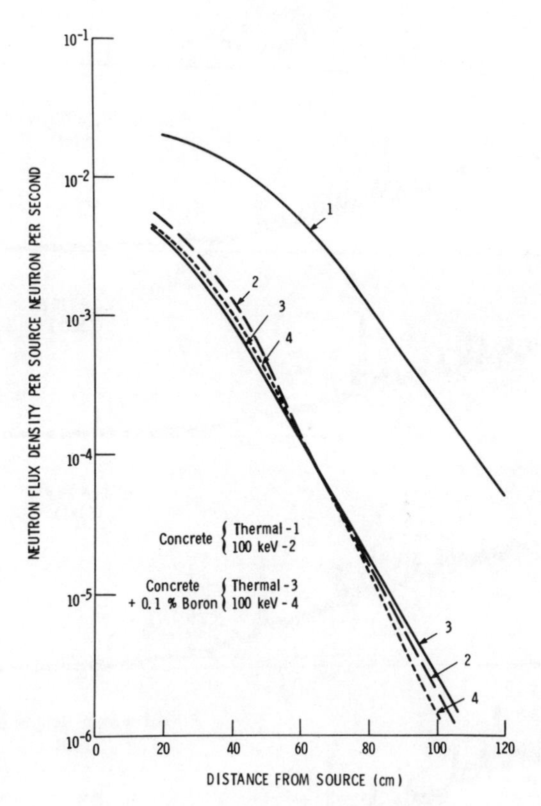

FIG.9. Theoretical calculation of the thermal (T) and epithermal (E) neutron flux density in normal and borated (0.1%) concrete.

is due primarily to thermal neutron capture. However, when using a fission neutron source this may not always be so. If the ratio of thermal- to epithermal-neutron flux density changes, it is conceivable that epithermal-neutron capture may become significant and alter the relative peak intensities.

Monte Carlo computations are not available for salt water, so we cannot show this effect quantitatively. However, the effect may be illustrated by the change of neutron energy distribution brought about by adding boron to concrete. One-tenth percent of boron should have about the same effective cross section as the combined cross sections for chlorine and sodium in seawater. The change in neutron energy distribution in concrete caused by adding 0.1 percent of boron is shown in Figure 9, according to data reported by Makra and Vertes [10].
The equilibrium populations of thermal and 100-keV neutrons are about the same in the borated concrete, in contrast with normal concrete, in which thermal neutrons predominate. By analogy one would expect a similar reduction in the ratio of thermal- to epithermal-neutron flux density in seawater.

The shift toward higher average neutron energy in seawater is reduced in the sediments by two factors which contribute to the reduction of chlorinity: displacement of the seawater by the solid material (about 40 percent in these sediments); and strong biochemical activity which continuously produces fresh water. These two factors can be expected to vary with the density

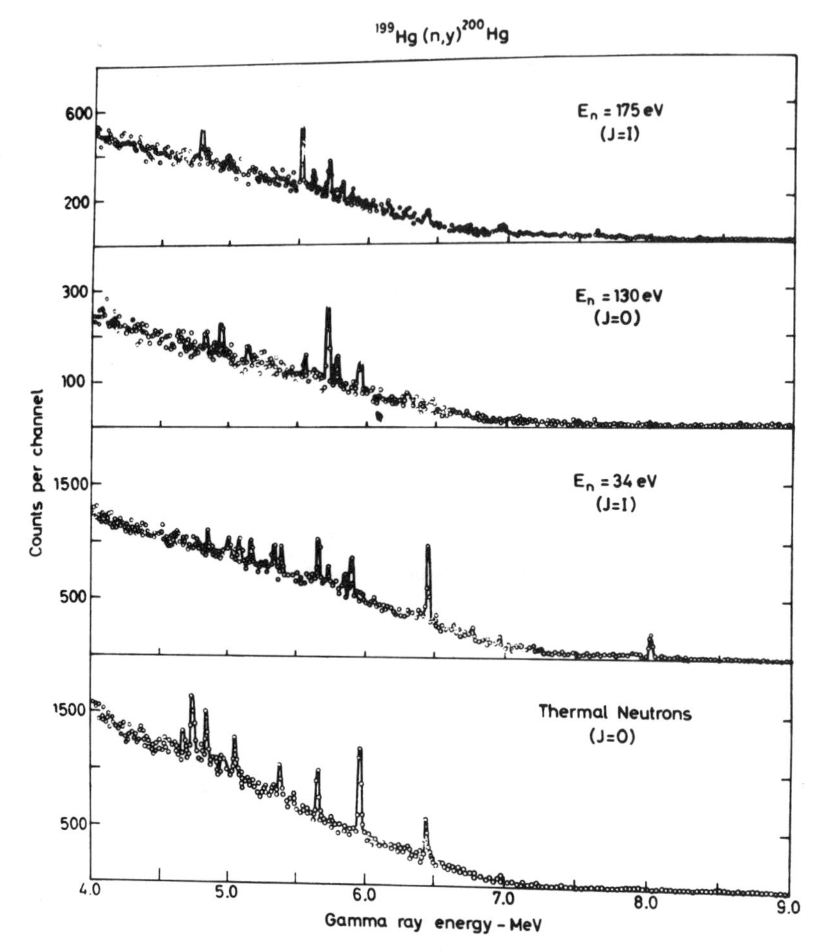

FIG.10. Thermal and resonance capture spectra of mercury at several neutron energies for the reaction $^{199}Hg(n,\gamma)^{200}Hg$ (from Ref. [11]).

and composition of the sediments, the amount of biochemical activity, and the depth of penetration of the instrument assembly. The higher the chlorinity and the lower the porosity (such as in a massive hard rock or coarse gravel), the greater the average neutron energy. Thus, the thermal- to epithermal neutron flux density can be expected to be somewhat different at each of the stations at which measurements were taken.

THERMAL- vs EPITHERMAL-NEUTRON CAPTURE GAMMA-RAY SPECTRA

Many elements undergo resonance neutron capture. Figure 10 shows the capture gamma-ray spectrum for the reaction ^{199}Hg (n,γ) ^{200}Hg as measured by Rae et al [11] for thermal and several specific epithermal energies. The spectral differences for the different neutron energies are obvious. Each neutron energy population generates a distinctive capture gamma-ray spectrum. In

TABLE III. COMPARISON OF THE INTENSITY OF SPECIFIC CAPTURE GAMMA-RAY LINE FROM ^{35}Cl AFTER CAPTURE BY THERMAL, 25-keV AND 50-keV NEUTRONS

Energy	Intensities (relative)		
$E\gamma$	I_{th}	I_{25}	I_{50}
8.579	3.0	1	1
7.790	8.4	1.2	6.1
7.414	12	0.6	1.4
6.978	2.5	0.2	2.0
6.620	13.3	0.8	2.3
6.111	23	0.3	4.5
1.957	15 }	~0.5	6.7
1.951	23 }		

Data from Ref.[12].

effect, our measurements represent a summation of many spectra from various epithermal-neutron populations in addition to the thermal-neutron-capture spectrum. One should not be surprised by marked differences among capture gamma-ray spectra depending upon the neutron energy distribution in the volume from which the gamma rays are detected. As the resonance bands are narrow and quite numerous for some nuclei, it should not take a large energy shift in the neutron distribution to cause a significant change in the resulting capture gamma-ray spectrum.

The differences in neutron-energy distribution caused by differing concentrations of hydrogen and chlorine may explain the inconsistencies of amplitude--or even of the presence or absence--of peaks in the capture gamma-ray spectra of target materials similar in composition. The average energy of the neutron flux density in the sediments is lower than that in seawater. The thermal-neutron-capture cross section of ^{35}Cl is about 42 barns, whereas the resonance integral is less than 20 barns for the same isotope. However, as shown in Table III the resonance absorption in the region of 25 to 50 keV yields peak intensities which are significantly different from those for thermal neutrons. The reduction or enhancement of the capture probability of intermediate energy neutrons by ^{35}Cl may account for some of variation observed in peak amplitude.

Similar arguments can be made for over 40 elements which have resonance integrals which are several orders of magnitude larger than their thermal cross sections.

LABORATORY SPECTRAL MEASUREMENTS

To test the above hypothesis, a laboratory experiment was made to ascertain what changes in peak amplitude might be expected. A thin iron can having a center well was filled with 3.85 kg of dry NaCl and mounted about 50 cm below the surface of a large pool of water 1 metre deep. A 66-μg ^{252}Cf source was placed at the bottom of the well in a position approximately at the center of the sample can. A Ge(Li) detector was suspended 75 cm from the source and at approximately the same depth as the ^{252}Cf source. Using that same geometry, spectra were recorded when the sample can was completely clad with a 0.38-mm-thick cadmium sheath and when it was not. The amplitudes of the 5.089-, 5.600-, and 6.111-MeV peaks were summed in each spectrum, and the ratio of the sums was used to normalize the spectrum of the unclad NaCl to the spectrum of the cadmium-clad NaCl.

TABLE IV. COMPARISON OF PEAK AMPLITUDES FOR UNCLAD AND CADMIUM-CLAD NaCl

Energy	Counts in Peak		
	Unclad NaCl	Cadmium-clad NaCl	Percent variation
1166	1827	1767	3
1597	568	3321	- 83
1952	2040	2082	- 2
1984	460	2876	- 84
2351	234	1053	- 77
2702	103	990	- 90
2802	205	719	- 71
2864	861	1048	- 18
3706	147	498	- 70
3789	77	932	- 92
3960	608	547	11
3983	124	485	- 74
4470	346	364	- 5
4497	282	649	- 57
4693	984	1106	- 11
4735	112	445	- 75
4945	131	366	- 64
4980	434	472	- 8
5089	3697	3748	- 1
5205	620	545	14
5247	148	677	- 78
5600	3598	3699	- 3
5716	492	512	- 4
5956	470	386	22
6111	2586	2435	6
6392	1899	1890	0
6467	257	303	- 15
6621	588	609	- 3
6769	1556	1505	3
6903	1176	1113	6
6978	151	184	- 18
7279	991	1035	- 4
7413	592	690	- 14
7557	459	493	- 7
7790	490	471	4
8067	293	311	- 6
8577	119	132	- 10

The amplitudes of the major chlorine peaks are shown in Table IV for the spectra of both the unclad and clad NaCl. The percent deviation is shown in the last column. An examination was made of each peak shape and possible interference for those peaks having a deviation of greater than 50 percent. The large deviation could be assigned either to interference from another element or to the effect on the computation algorithm of a peak distorted by statistical extremes over a few adjacent channels. Some peaks showed variations of 10-20 percent but appeared to be valid. Although not definitive, these variations are thought to be caused in part by differences in thermal and epithermal neutron populations at the points of capture.

An experiment was also made using crushed manganese nodules instead of NaCl. Similar results were obtained. The variations in peak intensity appeared to be greater than those expected from statistical variations in the number of counts in the respective peaks. Prestwich and Coté[13] have shown significant differences in the thermal capture and resonance spectra of manganese and the intensity variations observed in these experiments should be expected.

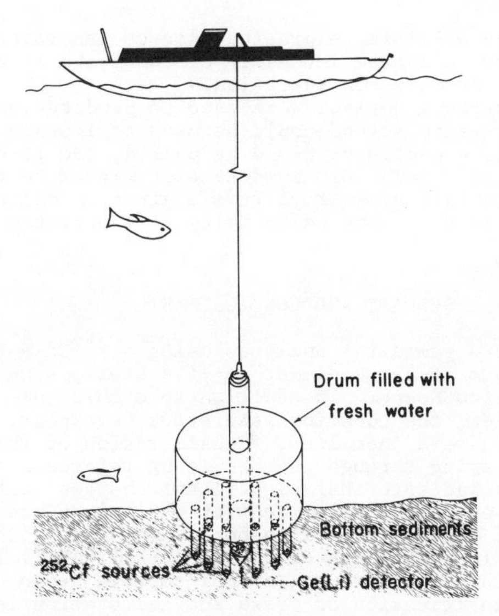

FIG.11. Sketch showing a proposed improved probe using a multiple ^{252}Cf source in a more efficient array.

RECOMMENDATIONS

These experiments raise some problems in the application of the capture gamma-ray method to seabed elemental analysis. The chlorine problem is more serious than previously expected, and from a practical point of view it does not seem possible to obtain good counting statistics with the geometry used in these experiments. A higher intensity source would help, but to obtain counting statistics that would give good quantitative data in a practical counting time a much larger source would be desirable. In this case, the life of the detector would probably be impaired. Finally, it must be recognized that the spectra observed are not true thermal capture spectra but are a summation of thermal, epithermal resonance-and slow neutron capture spectra. Depending on the element and the thermal-to-epithermal neutron ratio, good calibration may or may not be difficult to attain.
Some measures are suggested to reduce the problems.

(a) By changing the configuration of the probe, it may be possible to improve the counting statistics. It should be more efficient to place the detector in the nose, at the position of the source in our present probe. Instead of using a single rather strong ^{252}Cf source, it would be more advantageous to use weaker sources distributed about the detector and mounted in similar pointed probes that would penetrate the sediments to the same depth as the detector. A ring source constructed of ceremet wire could also be used. If the individual sources or wire are all placed 13 to 15 cm from the detector, the thermal and epithermal neutron flux density would be least affected by variations in density of the sediments but a larger source-to-detector distance may be necessary to protect the detector. Figure 11 shows a possible configuration.

(b) The neutron energy problem is one of calibration that will be a function of energy distribution of the neutron flux density through the sample. By taking the ratio of two peaks in the spectrum of some common element such as iron in the sample, an index of spectral hardness might be established. From such an index, it may be possible to predict through laboratory experiments what the spectral intensity of the peaks of a given element should be per unit concentration.

(c) If a neutron generator were used to produce 14 MeV neutrons, an alternative method could be used to improve the spectra. The neutron generator could be pulsed, and it should be possible by proper timing of the detection window to time the system to detect capture gamma rays from a given resonance spectrum and thus increase the selectivity of the system for a given element.

SUMMARY AND CONCLUSIONS

In situ capture gamma-ray analyses using a ^{252}Cf - Ge(Li) probe have been made of the sediment at five stations on the bottom of Long Island Sound. In addition to a chlorine interference problem, the counting statistics were poor, and the peak amplitudes were anomalous. Consideration of the energy of the neutrons passing through the sample by reference to Monte Carlo calculations indicate that significant changes in the ratio of the thermal to epithermal flux density may take place and that the mean energy of the epithermal component may vary with the density of the sediments. Such changes, though relatively small, can cause large changes in the epithermal capture gamma-ray spectra. Thus, positive identification of peaks and calibration is difficult. These can be serious problems if the capture gamma ray method is to be used for seabed analyses. Some recommendations for future experiments are presented.

ACKNOWLEDGEMENTS

The authors are indebted to Mr. Joseph Vadus of the National Oceanic and Atmospheric Administration who organized the several agencies involved in the experiments and to the personnel of the International Underwater Contractors for their help in carrying out the work. We are also grateful for the assistance of three guest investigators Dr. Hans Fanger, Gesellschaft für Kernenergieverwertung, West Germany, Dr. Paul Huppert, Commonwealth Scientific and Industrial Research Organization, Australia, and Mr. Ronald Wallace, University of Georgia. The ^{252}Cf source was loaned by the U. S. Atomic Energy Commission and partial support was received from the National Sea Grants Program, U. S. Department of Commerce.

REFERENCES

[1] Noakes, J. E., and Harding, J. L. (1971) New techniques in seafloor mineral exploration. Marine Tech. Soc. Jour. 5, pp 41-44.

[2] Moxham, R. M., Tanner, A. B., and Senftle, F. E. (1975) *In situ* neutron activation analysis of bottom sediments, Anacostia River, D.C., Geophysics (in press).

[3] Cooper, J. A., Nielson, H. L. and Perkins, R. W. (1973) Feasibility study of *in situ* sediment analysis by X-ray fluorescence, Battelle Pacific Northwest Laboratory Rept. BNWL-SA-4814, 15 pp.

[4] Senftle, F. E., Duffey, D. and Wiggins, P. F. (1969) Mineral exploration of the ocean floor by *in situ* neutron absorption using a californium-252 (^{252}Cf) source, Jour 3, 9-16.

[5] Wiggins, P. F., Senftle, F. E. and Duffey, D. (1970) Neutron capture gamma-ray analysis of marine manganese nodules using ^{252}Cf, 13, 60-63.

[6] Senftle, F. E., Evans, A. G., Duffey, D. and Wiggins, P. F. (1971) Construction materials for neutron capture gamma ray measurement assembly using californium-252, NUCL. TECH. 10, 204-210.
[7] Boynton, G. R. (1975) Canister cryogenic system for cooling germanium semiconductor detector in borehole and marine probes, (in press).
[8] Tanner, A. B. (1975) A sliding digital filter for rapid reduction of nuclear pulse-height spectra, U. S. Geol. Survey, open file report.
[9] Stoddard, D. H. (1974) written communication.
[10] Makra, S., and Vértes, P. (1973) Spectra of neutrons transmitted through shields of different materials and thicknesses, Kernenergie 16 (12), 378-383.
[11] Rae, E. R., Moyer, W. R., Fullwood, R. R., and Andrews, J. L. (1967) Gamma-ray spectra from radiative capture of thermal and resonance neutrons in mercury and tungsten, Phys. Rev. 155, 1301-08.
[12] Bird, J. R., Bergquist, J., Biggerstaff, J. A., Gibbon, J. H., Good, W. M. (1973), Compilation of keV neutron capture gamma ray spectra, Nucl. Data Tables 11(6), 433-529.
[13] Prestwich, W. V. and Coté, R. E. (1967) Gamma-ray spectra of Co^{60} and Mu^{56} following resonance-neutron capture in Co^{59} and Mu^{55}, Phys. Rev. 155, 1223-1229.

COMPARISON OF NUCLEAR WELL LOGGING DATA WITH THE RESULTS OF CORE ANALYSIS

The significance of nuclear well logging in geological exploration

J. A. CZUBEK
Institute of Nuclear Physics,
Cracow, Poland

Abstract

COMPARISON OF NUCLEAR WELL LOGGING DATA WITH THE RESULTS OF CORE ANALYSIS: THE SIGNIFICANCE OF NUCLEAR WELL LOGGING IN GEOLOGICAL EXPLORATION.

It is possible to consider the results of a geophysical assay of a formation statistically. In this case the geological formation in question is considered as a field for the regionalized variable. The regionalized variable itself (in the sense of Matheron's theory of geostatistics) is defined on some geometrical support being equal to the volume of the sample. In this case the results, or rather their scattering around the mean value, depend on the volume of the sample. This leads to some discrepancies between the results of the geological assay (on the core samples) and the geophysical data. By using the concept of the absolute dispersion coefficient in the de Wijs scheme, it is possible to obtain a new approach to the problem of how to construct the calibration curves for the nuclear probes on the basis of the field borehole logging and the laboratory measurements on the core samples. The method seems to be very interesting for many practical purposes in the evaluation of geological formations for oil and for solid mineral exploration.

INTRODUCTION

Geophysical well logging data are usually exploited for two main purposes: for qualitative geological recognition of formations, and for quantitative evaluation of such parameters as porosity, bulk density, ore grade, shale content and oil saturation of the formations being considered. Among the logging methods, the nuclear ones are the most important, especially for the quantitative evaluations.

The usual procedure in these quantitative nuclear logging methods is the calibration one which can be performed in two different ways:

(1) The probe response versus the parameter of interest (porosity, density, ore grade, etc.) is obtained on model assemblies consisting of the well-known, very homogeneous rock materials.

(2) The calibration curve is obtained in the field by comparing the probe responses at given points in the boreholes with results of the laboratory analysis of core samples collected at the same points.

The other calibration procedures are less important and are used only occasionally.

Results of the quantitative interpretation of nuclear logs are always statistically checked by the core sample analysis at some points. And it is here that the discrepancies arise: the check points for results obtained by the first calibration procedure are very often in disagreement with the core sample analysis, whereas for the second kind of calibration procedure this discrepancy is observed when the same calibration curve is used in two different geological regions (even when the lithology is the same), or when the checking procedure is performed by a different geophysical company than the one which performed the probe calibration.

Even if one disregards the trivial sources of these discrepancies like incorrect standardization of the probe readings, troubles in operation of the probe electronics, errors in fitting the proper depth to the log and laboratory data, etc., it is still difficult to assess the reason for these often wide differences between the geophysical and laboratory data — the latter usually being considered correct.

This paper tries to explain and even justify the situation and indicate how to avoid it. The problem seems to be one of great importance, especially in those cases where the new nuclear logging technique has been introduced and checked against the old conventional methods, and also when a very precise calibration curve is required for the nuclear probes.

WHAT IS ACTUALLY MEASURED IN THE FIELD?

To the geologist, the obvious reason for these discrepancies is the differences in the objects being measured: the geophysical data are collected from rather large volumes (of the order of 10^5 to 10^7 cm^3) of rock taken from outside the borehole, whereas the laboratory data are usually obtained from small samples (of the order of 20 to 100 cm^3) of the core taken from the inside of the borehole. Thus, the first question to anwer is to what extent does the <u>volume</u> of the rock sample influence the result of the measurement. To be able to do this we shall use the geostatistical approach following the excellent work of Matheron [1-3].

When a rock matrix is considered, different models are possible, but above all we must keep in mind that although it may appear to be homogeneous it is in reality heterogeneous. This is especially true when such rock parameters as porosity, bulk density and ore grade are considered. To simplify our reasoning (but without any loss of its generality) we can consider, for example, the porosity.

If for some geological formation of volume V we wish to know how large the scattering of the measured porosity is as a function of the volume v of the considered sample, the situation will be as that sketched in Fig.1. For the sample volume equal to zero (or almost negligible in comparison with the rock grain sizes) our sample will be either in the rock grain (porosity 0) or in the pore (porosity 1.0) which are given by the points A and B, respectively, in Fig.1. The frequency with which the result of the measurement will be A or B depends on the grain size distribution and their space distribution, and for this reason is quite random.

The second extreme case is when volume v of the sample is equal to volume V of the whole formation. Here the result of the measurement has nothing in common with randomness (apart from the usual random measurement error); it is the well-defined average porosity of the whole system. This is given by point C in Fig.1. In the triangle ABC in Fig.1, the notion of porosity loses its random meaning going from the side AB toward the apex C. Thus, the porosities of the samples collected randomly in the

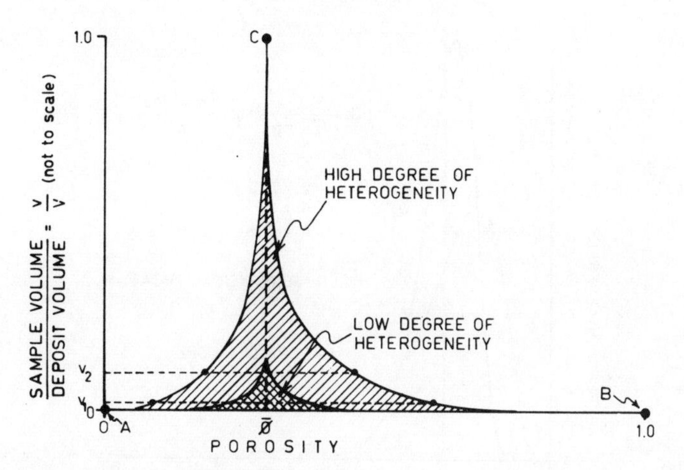

FIG. 1. Range of the observed porosity values scattered around the mean porosity $\overline{\phi}$ for different volumes of rock samples and for two kinds of deposit heterogeneity.

formation, as a function of their sizes v, will be scattered around the average value $\overline{\phi}$, as shown in Fig.1 by the shaded area. The 'convergence' of the measurement results toward the average $\overline{\phi}$ value, as a function of the sample size v, will be higher the lower the heterogeneity (for the moment without its definition) of the formation. This situation is shown in Fig.1 by the double-shaded area. The points A and B in Fig.1 are purely artificial from the practical point of view (one never collects such small samples), whereas the point C is not. This is the situation when the whole deposit (i.e. the volume V) is still exploited and the results of such exploitation (the total amount of mineral, etc.) are known. Thus, unfortunately 'a posteriori', the position of point C is available. The usual situation, however, is the one given by the broken lines in Fig.1: small samples, each of volume v_1, are taken from the core, whereas the volume $v_2 > v_1$ is considered in each geophysical measurement. Conformable to this situation, the distributions of experimental data, say geophysical and laboratory type, will be different, the first being more and the second being less centralized around the mean value $\overline{\phi}$. This is shown in Fig.2.

In Fig.2 the log-normal distribution[1] of porosity in the deposit was assumed with different variances σ_G^2 and σ_L^2 for the geophysical and laboratory data, respectively. Here, the distributions of the geophysical and laboratory data not being symmetrical around the mean porosity value, the modal and the median values are different in each distribution (the mean value being the same). In consequence, the cumulative distribution functions of these densities will also be different and asymmetrical (Fig.3).

Let us assume, for a moment, that each pair of 'geophysical' and 'laboratory' samples has the same common geometric middle point in the space inside the geological formation being considered. All samples being distributed in this space, either randomly or regularly, can be considered as the bi-variate distribution of the geophysical ϕ_G and laboratory ϕ_L

[1] The log-normal distribution is a Gaussian distribution in logarithmic scale.

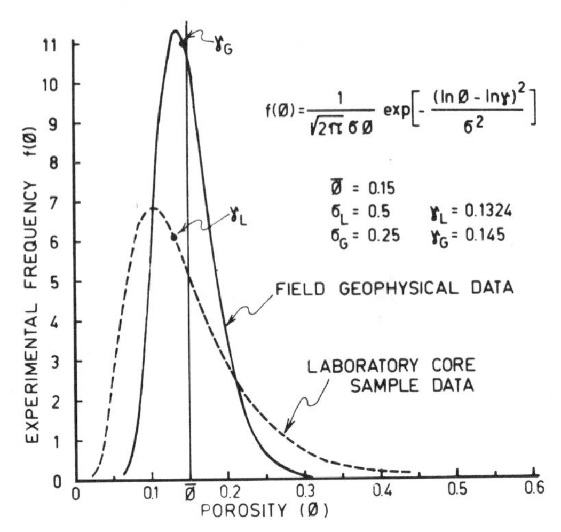

FIG. 2. Influence of the sampling volume on the shape of the experimental data distribution obtained from the same geological formation.

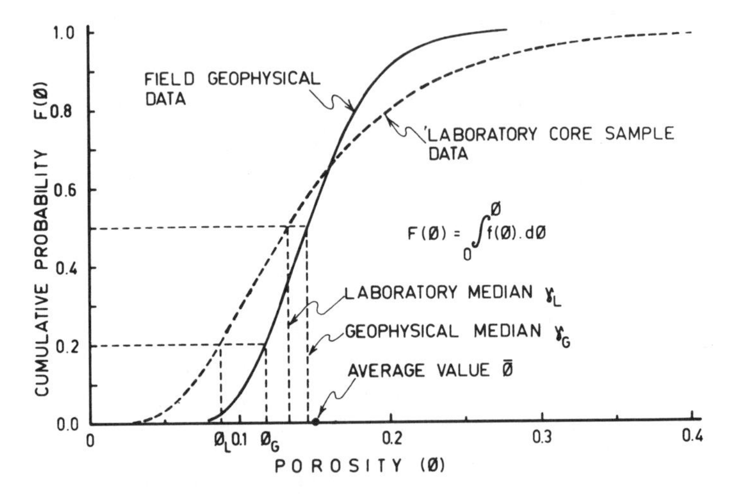

FIG. 3. Cumulative probability of geophysically determined and laboratory determined porosities of the same deposit (log-normal distribution).

random variable (in the statistical sense). In consequence, the probability density functions given in Fig.2 can be considered as the marginal distributions of this bi-variate distribution. Thus, the question of the correlation between the ϕ_G and ϕ_L data arises. From the considerations given above we can immediately say that, as the variance of laboratory data is greater than that of the geophysical data, the slope $d\hat{\phi}_G/d\hat{\phi}_L$ of the regression curve ($\hat{\phi}_G$ being the most probable value for a given $\hat{\phi}_L$ value) will always be lower than 1.0. Next, if the marginal distributions are not symmetric ones the regression line will not be a straight one. Thus, the regression line in this

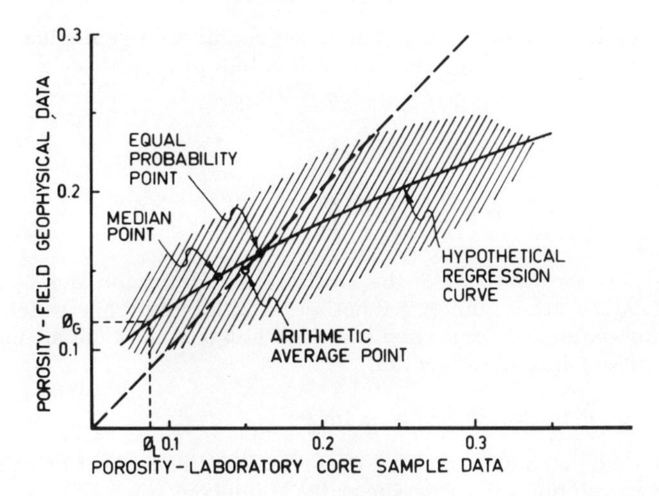

FIG. 4. Correlation between the laboratory and geophysical assay of the deposit porosity (log-normal distribution).

case will be similar to the one given in Fig.4. The shaded area in this figure represents in a schematic way the "cloud" of experimental data. The situation given in this figure occurs when the source of information about the 'true' porosity needed for the interpretation of the field geophysical data is different from the core samples — for example the calibration curve of the 'first kind' (obtained on the nearly homogeneous model assemblies). When the data for the laboratory core sample are used to construct the calibration curve (the 'second kind' of calibration procedure), the regression line in Fig.4 is transformed 'by force' to the straight, bisecting line. In this case, however, when one wants to use such a calibration curve to evaluate the formation in another geological region, the data obtained can be wrong owing to the simple fact of a different degree of heterogeneity of this other deposit.

As the main result of the foregoing consideration we arrive at the general conclusion that each geological data assembly (collected from some formation) is self-compatible only when all data are measured on a set of samples each having the same volume. If the same geological parameter of some formation is obtained on the two sample sets characterized by the two different sample sizes v_1 and v_2, the resulting parameter distributions are (or can be) different. A proper intercomparison between these two experimental distributions is only possible after some particular normalization in order to obtain the same statistical representativity for each set of samples. How this is achieved is explained in the next section.

SIMPLE QUANTITATIVE DESCRIPTION

To describe mathematically the qualitative result obtained above, a simple model of the rock heterogeneity is needed. Following the idea of Matheron [1-3], we shall use the notion of the semi-variogram $\gamma(\vec{r})$ to describe the half of the covariance of the considered geological parameter $P(\vec{r}')$ (porosity, ore grade, etc.) for two points at a distance $\vec{r}$ from each other, i.e.

$$\gamma(\vec{r}) = (1/2)\cdot D[P(\vec{r}' + \vec{r}) - P(\vec{r}')]$$

$$= (1/2)\cdot E[P(\vec{r}' + \vec{r}) - P(\vec{r}')]^2 \qquad (1)$$

where D(..) is the variance of the statistical distribution and E(..) is the expected value. Here, the $\gamma(\vec{r})$ function is an intrinsic one which characterizes the geological formation itself. When one uses the so-called de Wijs scheme of intrinsic function

$$\gamma(\vec{r}) = \gamma(|\vec{r}|) = 3\alpha \log_e|\vec{r}| = 3\alpha \log_e r \qquad (2)$$

the variance σ_v^2 of the parameter P determined from a set of samples each having volume v (as was shown by Matheron) is:

$$\sigma_v^2 = \alpha \log_e(V/v) \qquad (3)$$

where V is the volume of the whole formation being considered, and the coefficient α is the absolute dispersion: it characterizes the dispersion of the pure state phenomenon, that is, it is the measure of the formation heterogeneity. From Eq.(3) it follows simply that for two sets of results obtained on the v_1 and v_2 volume samples there is the relation:

$$\sigma_{v_1}^2 - \sigma_{v_2}^2 = \alpha \log_e(v_2/v_1) \qquad (4)$$

which is the formula we are seeking.

The direct application of Eq.(4) is rather difficult owing to the assumption existing in its derivation that volumes v_1 and v_2 are geometrically similar (that is $v_1 = \lambda^3 \cdot v_2$, λ being constant). To avoid this difficulty, Matheron introduced the idea of the linear equivalents 1* of a given volume v. This gives, instead of Eq.(4), the relation

$$\sigma_{v_1}^2 - \sigma_{v_2}^2 = 3\alpha \log_e(1_2^*/1_1^*) \qquad (5)$$

The linear equivalents of different geometrical forms of v are the following:

for a parallelepiped of sides a > b > c:

$$a + b + (c/2) \leq 1^* \leq a + b + c \qquad (6)$$

for a cube of side a:

$$1^* = 2.7\ a \qquad (7)$$

and for a cylinder of height h and radius R (h ≥ 2R) is:

$$\begin{aligned} 1^* &= h + 2.85\ R & 0 \leq R/h \leq 0.2 \\ 1^* &= 0.960\ h + 3.03\ R & 0.2 \leq R/h \leq 0.5 \end{aligned} \qquad (8)$$

as was found by Lenda [4].

The absolute dispersion coefficient α depends on the geological formation and, being always lower than 1.0, is of the order (for the log-normal distribution) of 10^{-3} to 10^{-4} for the sedimentary stratiform formation up to over 10^{-1} for the metallogenic igneous deposits. The methods of determining this coefficient are discussed in the next section.

Knowing the α value for a given deposit, let us consider an example. Returning to our porosity discussion, we can assume the core laboratory sample data set and the field neutron data set. The rock samples for porosity determination in the laboratory are cylinders of diameter $2R = 28$ mm and height $h = 30$ mm. The linear equivalent l_1^* of this volume, Eq.(8), is $l_1^* = 7.12$ cm.

The neutron-derived porosity is determined (using the calibration curve) from the average neutron reading on a section of 1.0 m. Taking the average neutron migration length in the porosity diapason of 5 to 30 per cent as being 20 cm, the volume v_2 of the "geophysical" sample can be considered as a cylinder of diameter 40 cm and height 100 cm (we disregard here the influence of the borehole). The linear equivalent l_2^* of such a sample is $l_2^* = 157$ cm. Thus, according to Eq.(5), we have:

$$\sigma_L^2 - \sigma_G^2 = 3\alpha \log_e(157/7.12) = 3\alpha \log_e 22.05 = 9.13\,\alpha$$

An example of the practical behaviour of the results furnished by Eq.(5) is given in Fig.5 where, for the assumed log-normal distribution of porosity, the values $\bar{\phi} = 0.15$, $\sigma = 0.25$ for $v_1 = 10^7$ cm^3 were taken as departure points for some deposit with the absolute dispersion coefficient $\alpha = 0.01$. Next, the calculation was repeated for some other deposit with the same $\bar{\phi}$ but with $\alpha = 0.001$. The $\pm 3\,\sigma$ belts are given in this figure for both deposits as a function of the variable volume of the samples. Now, the question is, how to find α.

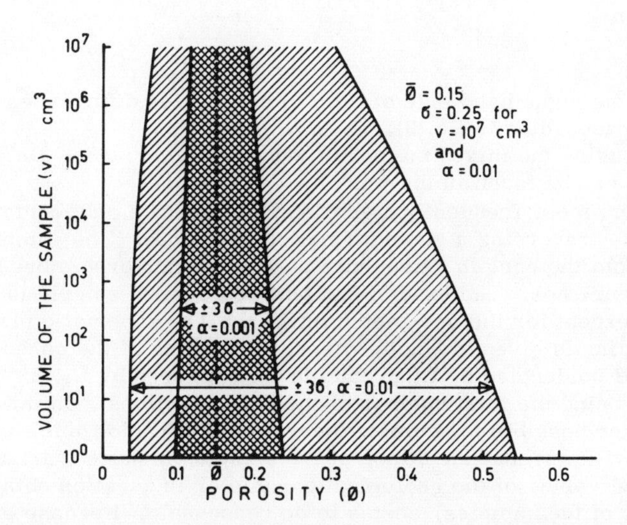

FIG. 5. $\pm 3\sigma$ belt around the mean value $\bar{\phi}$ for different sample volumes v and deposit heterogeneities α (log-normal distribution).

HOW TO FIND THE ABSOLUTE DISPERSION COEFFICIENT α

This is the most difficult theoretical and practical problem. The coefficient α, being itself a regionalized variable, can be determined only as an expected value $E(\alpha)$. For the regular grid of samples, the problem is still fairly simple. Matheron [1] presents three different possibilities:

(1) By using formula 3, where the volumes v and V are transformed into their linear equivalents:

$$\sigma_v^2 = 3\alpha \log_e(L^*/l^*) \tag{3a}$$

By knowing σ_v^2 from experiments and knowing the L^* and l^* values, the α coefficient is easily obtained from Eq.(3a).

(2) By calculating the slope of the variogram

$$\begin{aligned}\sigma_v^2(d) &= 6\alpha \log_e(d/l^*) + 9\alpha \\ &= 6\alpha \log_e d - 6\alpha \log_e l^* + 9\alpha\end{aligned} \tag{9}$$

where l^* is the linear equivalent of the sample volume v. The samples are at a distance d, and the covariance $\sigma_v^2(d)$ is calculated from the experimental values of the parameter P taken at the points $k \cdot L^\circ$ ($k = 0, 1, 2, .., n$), L° being the constant interval between the neighbouring samples. For $d = k \cdot L^\circ$ one calculates

$$\begin{aligned}s_v^2(d) &= s_v^2(k \cdot L^\circ) \\ &= \frac{1}{n-k}\sum_{i=0}^{n-k} \{P[(i+k)\cdot L^\circ] - P(i \cdot L^\circ)\}^2 \xrightarrow[n\to\infty]{} \sigma_v^2(d)\end{aligned} \tag{10}$$

The plot of the right-hand side of Eq.(10) against $\log_e d$ furnishes a slope equal to $6\,\alpha$ according to Eq.(9).

(3) By using the mixed method of variogram/variance. Details of this method can be found in the work by Matheron [1].

However, most frequently in geophysical practice only certain sections of a borehole traversing a given formation are cored. The samples are collected from the core in a sequence of one, two, or even more from each one-metre core box. The exact depth position of each sample is usually not known, except for the one-metre interval. The schematic plot of the sample spacing is given in Fig.6. Here, the lengths of the cored borehole sections will be denoted by L_{2r+1}, and those not cored by L_{2r} ($r = 0, 1, 2, ..., n$, $n + 1$ being the total number of the cored sections). In this situation it is no longer possible to evaluate theoretically a variogram similar to that given by Eq.(9), but the notion of the sample variance (variance of the experimental values of the geological parameter in question obtained from the analysis of the samples) seems to be reasonable. Because of the lack of information about the exact position of the sample inside a given core box, we must assume an equidistant sample distribution with the step L°.

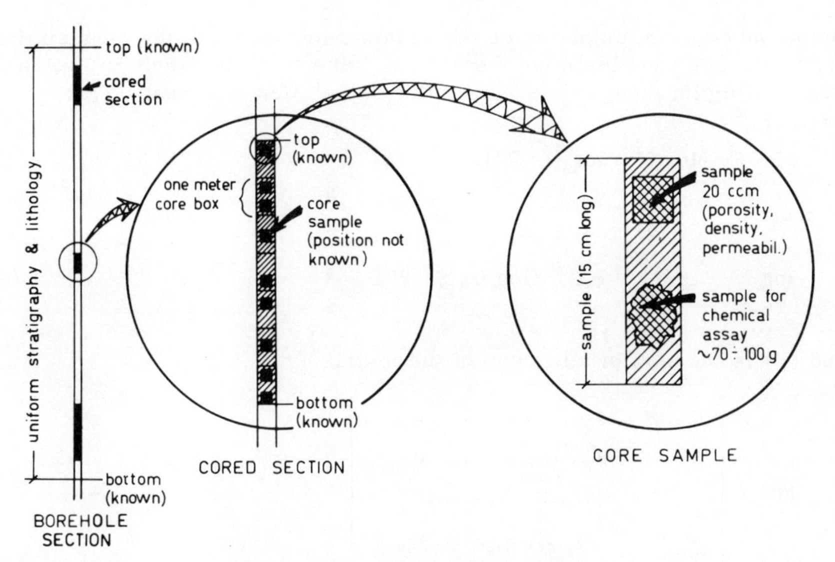

FIG. 6. Space distribution of core samples.

Thus, the number of samples in the section L_{2r+1} is

$$n_{2r+1} = \frac{L_{2r+1}}{L^\circ} \quad (r = 0, 1, 2, .., n) \tag{11}$$

whereas the number of the "lacking" samples in section L_{2r} is

$$n_{2r} = \frac{L_{2r}}{L^\circ} \quad (r = 1, 2, .., n) \tag{11a}$$

and the total number of samples is

$$N = \sum_{r=0}^{n} n_{2r+1} \tag{12}$$

Let each sample have an equivalent length l^* and the geological parameter P_i is determined in the sample of the consecutive number i (i = 1, 2, .., N). Then the experimental sample variance is

$$s_P^2 = \frac{1}{N-1} \sum_{i=1}^{N} (P_i - \overline{P})^2 \tag{13}$$

where

$$\overline{P} = \frac{1}{N} \sum_{i=1}^{N} P_i \tag{14}$$

On the other hand, using the de Wijs-Matheron concept of the geostatistical approach, one can obtain the theoretical value of σ_p^2 similarly to Eq.(3a). After a simple, though time-consuming calculation, one has:

$$\sigma_p^2 = 3\alpha \{\log_e N' + \log_e(L^\circ/1^*)\} \tag{15}$$

where

$$\log_e N' = \frac{1}{N^2} \sum_{i=0}^{n} (n_{2i+1})^2 \cdot \log_e n_{2i+1} + [T] \tag{16}$$

and [T] is the sum of all terms of the matrix T:

$$[T] = \left| \begin{matrix} a_{01}, a_{02}, a_{03}, \ldots, a_{0k}, \ldots, a_{0n} \\ a_{12}, a_{13}, \ldots, a_{1k}, \ldots, a_{1n} \\ \ldots\ldots\ldots\ldots\ldots\ldots\ldots \\ \ldots, a_{rk}, \ldots, a_{rn} \\ \ldots\ldots\ldots \\ a_{n-1,n} \end{matrix} \right|$$

$$= \sum_{k=1}^{n} a_{0k} + \sum_{k=2}^{n} a_{1k} + \sum_{k=3}^{n} a_{2k} + \ldots + \sum_{k=r+1}^{n} a_{rk} + \ldots + \ldots + a_{n-1,n} \tag{17}$$

and

$$\begin{aligned} a_{rk} \cdot N^2 = {} & |n_{(2k)} - n_{(2r+1)}|^2 \cdot \log_e |n_{(2k)} - n_{(2r+1)}| \\ & + |n_{(2k+1)} - n_{(2r)}|^2 \cdot \log_e |n_{(2k+1)} - n_{(2r)}| \\ & - |n_{(2k+1)} - n_{(2r+1)}|^2 \cdot \log_e |n_{(2k+1)} - n_{(2r+1)}| \\ & - |n_{(2k)} - n_{(2r)}|^2 \cdot \log_e |n_{(2k)} - n_{(2r)}| \end{aligned} \tag{18}$$

with

$$n_{(2r)} = \sum_{i=0}^{r} (n_{2i-1} + n_{2i}) \tag{19}$$

$$n_{(2r+1)} = \sum_{i=0}^{r} (n_{(2i+1)} + n_{2i}) \tag{19a}$$

Plotting the experimental values of σ_p^2 from Eq.(13) against the $\log_e N'$ from Eq.(16) for the K (K = 1, 2, ..., K) boreholes drilled in the region, the slope of this straight line, according to Eq.(15), should be equal to 3α. In practice, however, the α value being the regionalized variable, the experimental values of σ_P^2 (Eq.(13)) obtained for a small number of samples (low value of N) are of low accuracy. It is more convenient to plot the values

$$\frac{s_P^2}{\log_e N' + \log_e(L^\circ/1^*)} = \frac{s_P^2}{U} \xrightarrow[N\to\infty]{} 3\alpha \qquad (15a)$$

against the N values to obtain a good estimation of α for large values of N. The meaning of the symbol U is quite clear from Eqs (15a) and (15).

By knowing how to normalize the laboratory data to the geophysical ones it should also be possible to construct the proper calibration curves.

HOW TO TAKE ADVANTAGE OF STATISTICS FOR DETERMINING THE CALIBRATION CURVES

Let us assume that the nuclear probe readings are free from the influence of such parameters as

logging speed,
ratemeter time constant,
variable borehole diameter,
variable borehole fluid conditions,
presence or absence of casing tubes,
variable activities of possible radiation sources applied in the borehole tools,

and that we are limited to the one lithological type of formation being considered. Such normalized logging data are usually furnished by the interpretation procedures of different types, which are beyond the present discussion. The problem is how to translate these normalized nuclear logging data into the geological parameters being investigated, such as the shale content, bulk density, porosity and ore grade. This is possible, of course, when the proper calibration curve is available, but when it is not the preceding statistical discussion can help us to find it. This is, however, a rather difficult and complicated task, and we shall give here only a general outline of the method.

To make use of the method in question we should know the analytical relation between the normalized probe response and the parameter being investigated. For example:

(1) Natural gamma-ray intensity is a linear function of the uranium grade, or
(2) Natural gamma-ray intensity is a linear function of the volume of shale content, or
(3) The neutron-gamma intensity is a linear function of the logarithm of the "neutron" porosity, the neutron porosity being the weighted linear combination of the true porosity and the volume of shale content, etc.

In this sense, the geological parameters being statistically distributed in a given formation, the normalized probe responses also become the dependent random variables in this formation.

Besides these normalized nuclear logging data available from a certain number of boreholes, the geological parameters sought (porosity, shale content, ore grade, etc.) are also known from the laboratory analysis of the core samples collected from the boreholes (though not necessarily the same as for the geophysical data) drilled in the same formation. We recalculate the distribution of these laboratory data to normalize it to the volume of sample seen by the nuclear probe, as was shown in the foregoing discussion.

Now, using the standard statistical tests, we are able to know the type of this distribution (normal, log-normal, etc.), its expected value, variance, or even the higher moments.

From the functional relation, i.e. the normalized nuclear probe readings versus the geological parameter being considered, we can obtain, by the usual statistical procedure (see for example Papoulis [5]) the expected value, the variance and even the higher moments of the statistical distribution of this function. On the other hand, the same values are obtained directly from the logging data and a comparison of the theoretical with the experimental results is possible, this giving the elements necessary to construct the calibration curve. In practice we construct the confidence limits around the expected values of the two distributions, at the same time knowing the functional relations between them, which is still sufficient for drawing the calibration curve.

This procedure is illustrated by several simple examples.

Example 1

The normalized natural gamma-ray activity, I_γ, is proportional to the content of radioactive material in the formation. Under some conditions this radioactive material can be considered proportional to the volume of shale content V_{sh} in the rock. Thus, we have the relation

$$I_\gamma = a_1 + b_1 \cdot V_{sh} \tag{20}$$

where a_1 and b_1 are the unknown constants.

Let I_γ be normally distributed in this formation; this means that the average value $\overline{I}_\gamma$

$$\overline{I}_\gamma = \frac{1}{N} \sum_{i=1}^{N} I_{\gamma i} \tag{21}$$

is known, where $I_{\gamma i}$ are the normalized consecutive probe readings along the borehole axis. We know also the variance σI_γ^2 of this distribution as

$$s_{I_\gamma}^2 = \frac{1}{N-1} \sum_{i=1}^{N} (I_{\gamma i} - \overline{I}_\gamma)^2 \rightarrow \sigma_{I_\gamma}^2 \tag{22}$$

From the laboratory analysis of the core samples we obtain in the same manner the average $\overline{V}_{sh}$ value and the variance $\sigma_{V_{sh}}^2$. We recalculate this variance to a volume sample the same as that seen by the natural gamma-ray probe, say the $\sigma_{0, V_{sh}}^2$ value. From the behaviour of the function given by Eq.(20) we know that the $\overline{I}_\gamma$ value corresponds to the $\overline{V}_{sh}$ value and that

$$\sigma_{I_\gamma} = b_1 \cdot \sigma_{0, V_{sh}} \tag{23}$$

It is sufficient now to draw a straight line which passes through the point $\overline{V}_{sh}$, $\overline{I}_\gamma$ with a slope b_1 known from Eq.(23), and the calibration curve for the gamma-ray probe is ready.

Example 2

Let $I_{n\gamma}$ be the normalized reading of the porosity neutron-gamma probe (natural gamma-ray background subtracted). We assume the fundamental relation between the $I_{n\gamma}$ readings and the "neutron" porosity ϕ_n to be

$$I_{n\gamma} = a_2 - b_2 \cdot \log_e \phi_n \tag{24}$$

with the "neutron" porosity ϕ_n connected to the true porosity ϕ by

$$\phi_n = \phi + m \cdot V_{sh} \tag{25}$$

where m is a constant. We can take m = 0.40.

From the laboratory data we know the statistical distribution of ϕ and V_{sh} and it is possible to assume the log-normal distribution of the resulting ϕ_n data. In this case the statistical distribution of $I_{n\gamma}$ after Eq.(24) will be normal with the following parameters:

Expected value

$$\overline{I}_{n\gamma} = a_2 - b_2 \cdot \overline{\log_e \phi_n} \tag{26}$$

and the variance

$$\sigma_{I_{n\gamma}}^2 = b_2^2 \cdot \sigma_0^2 (\ln \phi_n) \tag{27}$$

where

$$\overline{\log_e \phi_n} = \frac{1}{N} \sum_{i=1}^{N} \log_e \phi_{ni} \tag{28}$$

and

$$\sigma^2 (\ln \phi_n) = \frac{1}{N-1} \sum_{i=1}^{N} (\log_e \phi_{ni} - \overline{\log_e \phi_n})^2 \tag{29}$$

and $\sigma_0^2(\ln \phi_n)$ is obtained from the $\sigma^2(\ln \phi_n)$ value after its normalization to the proper sample volume corresponding to the one "seen" by the neutron-gamma-ray probe. Here again, the calibration curve passes through the point $\overline{I}_n$, $\overline{\log_e \phi_n}$ with the slope b_2 known from Eq.(27).

The foregoing examples are the simplest ones, but a similar procedure can be applied for any assumed functional relation between the probe reading and the geological parameter being investigated. In each case the central point, corresponding to the expected values, is determined and the subsequent points corresponding to the different confidence limits around the expected values ($\pm\sigma$, $\pm 2\sigma$, etc.) are plotted, this giving the calibration curve. The standard deviations (the variances) on both axes should always be normalized to the same volume sample as that seen by the logging tool. Such a calibration curve has none of the inconveniences pointed out in the Introduction (which means that it is general). Furthermore, the troublesome search for the logging and laboratory data taken at exactly the same points is not needed here, because the marginal distributions of the bivariate distribution only are used. On the other hand, some more information about the character of the heterogeneous formation is indispensable.

CONCLUSIONS

The aim of this paper is to show how statistics can be applied to the evaluation of geophysical data. The paper explains:

(1) The possible origin of a discrepancy between the geophysical and laboratory data and how to avoid this discrepancy; and

(2) How to take advantage of statistics to construct the proper calibration curves needed in the quantitative interpretation of logging data.

REFERENCES

[1] MATHERON, G., Traité de géostatistique appliquée, Vol. 1, Mémoires du BRGM N° 14, Technip, Paris (1962).

[2] MATHERON, G., Les variables régionalisées et leur estimation, Masson et Cie., Paris (1965).

[3] MATHERON, G., The Theory of Regionalized Variables and its Applications. Les Cahiers du Centre de Morphologie Mathématique de Fontainebleau, N° 5. Publ. by the Ecole Nat. Sup. des Mines de Paris (1971).

[4] LENDA, A., Private communication, Inst. Nucl. Techn. of Mining-Metall. Academy, Cracow, Poland (1974).

[5] PAPOULIS, A., Probability, Random Variables and Stochastic Processes, McGraw-Hill, Inc., New York.

INSTRUMENTATION AND METHODOLOGY

SOME EXPERIENCE WITH THE USE OF NUCLEAR TECHNIQUES IN MINERAL EXPLORATION AND MINING

C.G. CLAYTON
UKAEA, Atomic Energy Research Establishment,
Harwell, Oxfordshire,
United Kingdom

Abstract

SOME EXPERIENCE WITH THE USE OF NUCLEAR TECHNIQUES IN MINERAL EXPLORATION AND MINING.

The paper describes briefly some of the work at Harwell in the areas of mineral exploration and mining, excluding uranium prospecting. Developments which are described include a towed sea-bed spectrometer which is finding important application in geological mapping of the sea-bed, analysis of geochemical samples and development of borehole logging equipment based on X-ray fluorescence techniques, coal ash monitoring and a Monte Carlo approach to evaluation of the spatial and energy distribution of neutrons in the media surrounding a borehole.

1. INTRODUCTION

At the present time there is considerable interest in the development and application of nuclear methods to resolve a wide variety of problems in mineral exploration and mining. The most important characteristics of these methods include a high specificity to elemental concentrations and this is often accompanied by an ability to operate satisfactorily in environments which often preclude the use of most other methods.

A number of techniques are currently being investigated for applications which vary from sea-bed and air-borne surveying, borehole logging, geochemical prospecting and mine control to on-stream analysis in mineral processing plants. Although much remains to be done, significant successes have already been demonstrated and new and improved instrumentation is continuously being introduced.

This paper is intended as a brief summary of some relevant areas of work in which we are currently involved.

2. RADIOMETRIC SURVEYING OF THE SEA-BED OF THE CONTINENTAL SHELF

Surface and air-borne radiometric surveys carried out over land during the past twenty-five years have shown that, in addition to locating radioactive mineral deposits, they are able to differentiate between the common types of rocks and sediments. As yet, however, few attempts have been made to extend these techniques to continuous measurements on the sea-bed and especially to the development of radiometric spectrometry capable of distinguishing areas of igneous and sedimentary rocks from measurements of their characteristic concentrations of uranium, thorium and potassium.

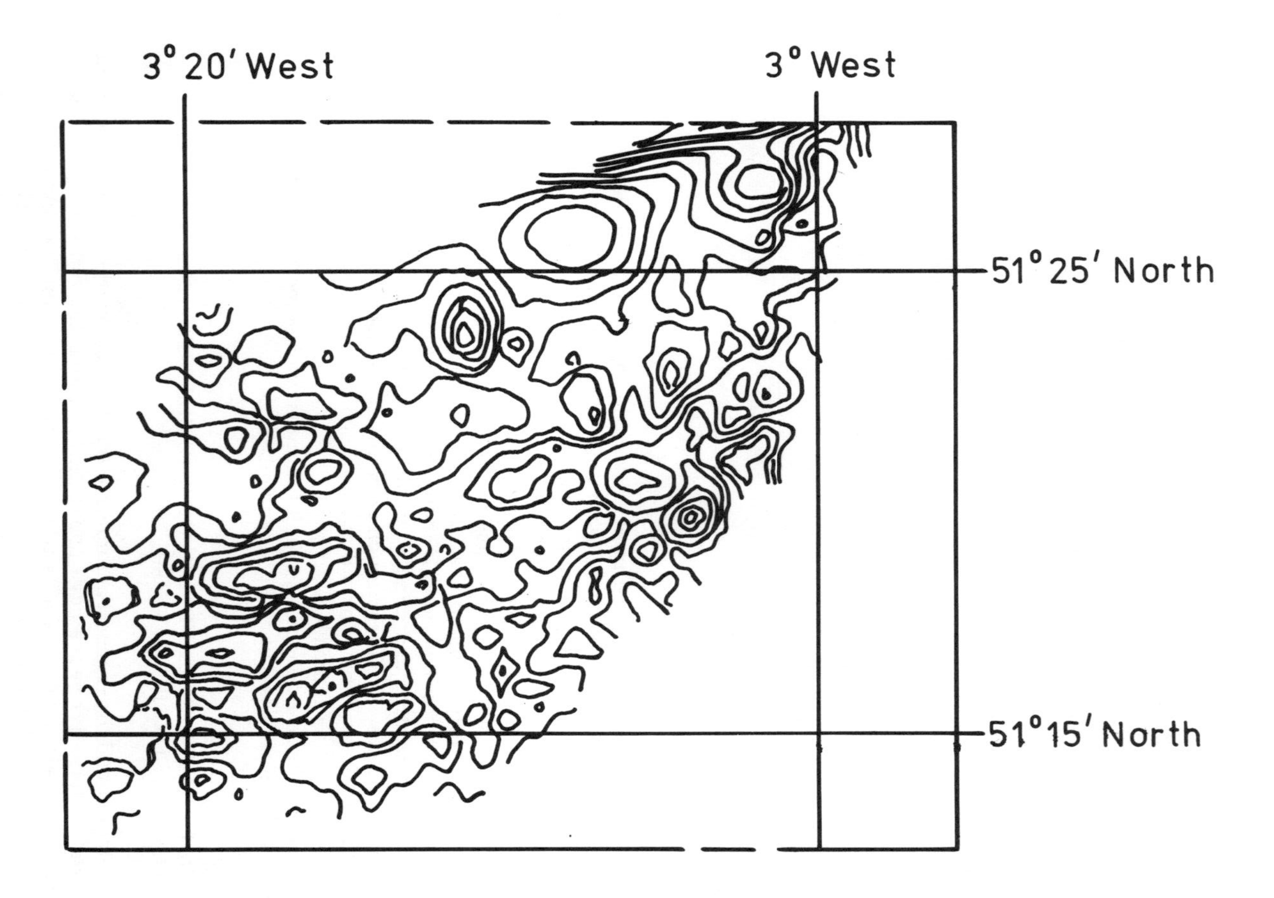

FIG.1. Contour map showing uranium activity in Bridgwater Bay [3]. Scale 1 in 2.5×10^5.

Apart from detecting the natural radioactive minerals, such equipment is also capable of detecting marine phosphate deposits and heavy mineral concentrations such as monazite, zircon and sphene which themselves contain more than average concentrations of radioactive minerals.

The towed sea-bed spectrometer developed at Harwell [1, 2] consists of a stainless-steel probe (12.7 cm dia. × 76.2 cm long) which contains a scintillation detector comprising a NaI(Tl) crystal (7.6 cm dia. × 12.7 cm long) and photomultiplier in an integrated assembly. The probe also contains the high voltage supply for the photomultiplier tube. The probe is towed continuously behind a ship and is connected to the ship-board electronic units by a double-armoured co-axial cable with an outer diameter of 1.2 cm.

To avoid the possibility of the probe becoming wedged between boulders, rock outcrops or wreckage on the sea-bed, the probe and the terminating 30-m length of cable are contained in a reinforced plastic tube or 'Eel'. The Eel is weighted so that the drag co-efficient of the Eel is the same as that of the cable: in this way a continuous curvature is maintained during towing. Under normal operating conditions about one-third of the length of the Eel rests on the sea-bed and the remainder is elevated to a height of approximately 7 m at the leading end. Such a configuration allows the Eel to clear any obstacles likely to be encountered on the sea-bed with the result that no equipment has been lost in over 10 000 km of surveying the sea-bed. The normal towing speed is between 4 and 5 knots, but over sedimentary deposits speeds of 7 knots have been maintained satisfactorily for long periods.

The equipment on-board ship comprises three main units: a winch of special design which can accommodate both cable and Eel, a four-channel spectrometer which registers total count-rate and count-rates in the energy bands corresponding to uranium, thorium and potassium activities, and digital and analogue recorders. Operation of the winch is controlled from the cabin in which the electronic units are mounted and this cabin also contains equipment which receives navigational data and sonar depth recorders. As the depth of the sea-bed is seen to be changing, the winch is operated in order to ensure that the correct length of cable is in use.

Before the probe reaches the sea-bed (when it is totally surrounded by sea-water), the total count-rate is approximately 7 counts/s. When the probe is in contact with the bottom, count-rates vary between 25 and 200 counts/s depending on rock type. At these count-rates a time constant of 4 s gives an acceptable error in the results and allows recognition of changes in the radioactivity of the sea-bed in a measuring time of 12 s. This corresponds to a distance of 25 m when the probe is traversing the sea-bed at a speed of 4 knots. The count-rates in the channels corresponding to potassium (1.5 to 5.0 counts/s), uranium (10 to 50 counts/min) and thorium (7 to 30 counts/min) are significantly smaller so that in order to gain statistically useful information, counting times of 500 s are normally used. This equates with approximately 1 km of traverse.

The equipment has now been developed to a stage at which it is being used semi-routinely for sea-bed surveys of the Continental Shelf around the United Kingdom.

The data from the spectrometer is currently recorded on paper tape, and computer programs have been developed which transform these data into a series of contour lines corresponding to iso-count curves for each of the naturally occurring radioactive elements and also for ratios of their radio-

TABLE I. LIST OF CASES STUDIED AND FLUX CONTOUR NUMBERING (Source: ^{252}Cf, 10^8 n/s)

Borehole radius (cm),	contents	Liner thickness (cm)	Formation (% water)
-	-	-	0
-	-	-	15
3	void	-	15
13	void	-	15
3	water	-	15
13	water	-	15
3	void	3	15
13	void	3	15
3	water	3	15
13	water	3	15
3	void	3	15 (boundary, 30-cm radius)
3	void	3	0
3	void	3	0 (boundary, 30-cm radius)

Contour No.	Flux ($n/cm^2 \cdot s$)
1	10^6
2	3×10^5
3	10^5
4	3×10^4
5	10^4
6	3×10^3
7	10^3
8	3×10^2
9	10^2
10	3×10
11	10

activities. A contour map showing the uranium activity in Bridgwater Bay in the British Channel is shown in Fig.1. Data of this kind are now proving to be a valuable aid in the construction of geological maps.

3. FORMATION ANALYSIS BY STEADY-STATE NEUTRON INTERACTION TECHNIQUES

Much of the past work on neutron transport has been carried out on relatively simple geometric systems, such as semi-infinite and infinite homogeneous media, because of the mathematical complexity involved when attempting complete solutions of the neutron transport equation. Published work in the field of formation logging uses various simplified mathematical approaches in basic geometries or else uses more complex approximations to complete analytical treatments of realistic borehole models.

To overcome the problems posed by a purely analytical solution of the complex situations to be found in a borehole, we have used a Monte Carlo program to study neutron behaviour in borehole conditions [4]. This program properly takes account of the type of neutron source, the neutron energy spectrum, the angular distribution of neutron scattering and various zones of different materials in the formation. For each neutron collision the type of interaction, the direction of flight, the subsequent energy and the mean free path to the next collision are determined from the range of possibilities taking account of available neutron cross-section data by random selection. In this connection the program itself generates a set of random numbers. For a particular number of neutron histories which are tracked, the average neutron flux within a given spatial region and energy band is proportional to the number of track lengths.

It is important to recognize that the greater the number of neutrons which are tracked, the greater the statistical accuracy of the results. Also, the statistical accuracy will be poorest where the flux is low: at distances remote from the source. However, it should also be recognized that if any spatial or energy regions are of special interest, the statistical accuracy can be increased to within any specified limits by tracking a sufficient number of neutrons: that is by expenditure of sufficient computing time.

The different borehole configurations examined so far are given in Table I and these apply to a formation consisting of pure silica, either dry or containing 15% by weight of water. Previous calculations have shown that whereas the presence of water appreciably affects the neutron distribution in the formation, the nature of the matrix material is, by comparison, not a major factor. As can be seen from Table I, starting from a point source in an infinite medium of dry silicon, a progressive study has been made of the effects of water in the formation, of two sizes of empty borehole (3 cm and 13 cm radius), of water in the boreholes, of the effect of a steel liner and of a finite radial boundary. The borehole sizes were selected to embrace situations of greatest interest in the metalliferous minerals and petroleum industries. Calculated flux distributions have been derived for thermal neutrons (zero to 0.1 eV), epithermal neutrons (0.1 eV to 1 keV) and fast neutrons (1 keV to 14 MeV) for a source of californium-252.

Only a few of the derived results are shown here: a more complete selection of results is given in the paper by Sanders et al. [4].

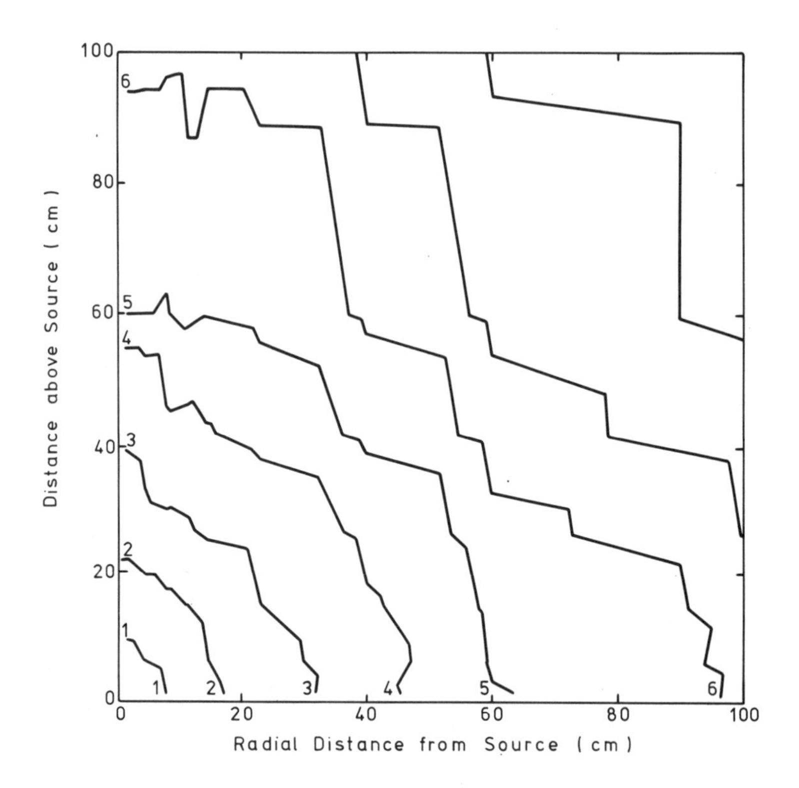

FIG. 2. Fast neutron flux distribution.
Formation: silica.
No borehole present.

The very significant effect of water in the pore space on the fast neutron flux distribution can be seen from Figs 2 and 3 which refer to a point source in an infinite medium of silica containing zero and 15% by weight of water. It can be seen, for example, that contour 5 (3×10^4 n/cm^2 · s) moves from about 64 cm radius to about 31 cm radius. An even greater change in flux gradient is seen for epithermal neutrons (Figs 4 and 5) and this reflects the distributed epithermal source (the fast neutron group) being changed as well as the epithermal neutrons themselves being more rapidly removed to the thermal group. The derived data also show that the thermal neutron flux gradient is also markedly increased throughout so that the spectrum remains "hard": that is it retains a high fast-to-thermal flux ratio. This is a consequence of the fact that while water is a good neutron moderator, it is also an appreciable thermal neutron absorber.

The effect of introducing a water-filled borehole of 13 cm dia., compared with an empty borehole, on the fast flux distribution can be seen by comparing Figs 6 and 7. As would be expected, the presence of a water-filled borehole results in a very significant increase in flux gradient within the borehole and this also strongly influences the flux gradients within the formation.

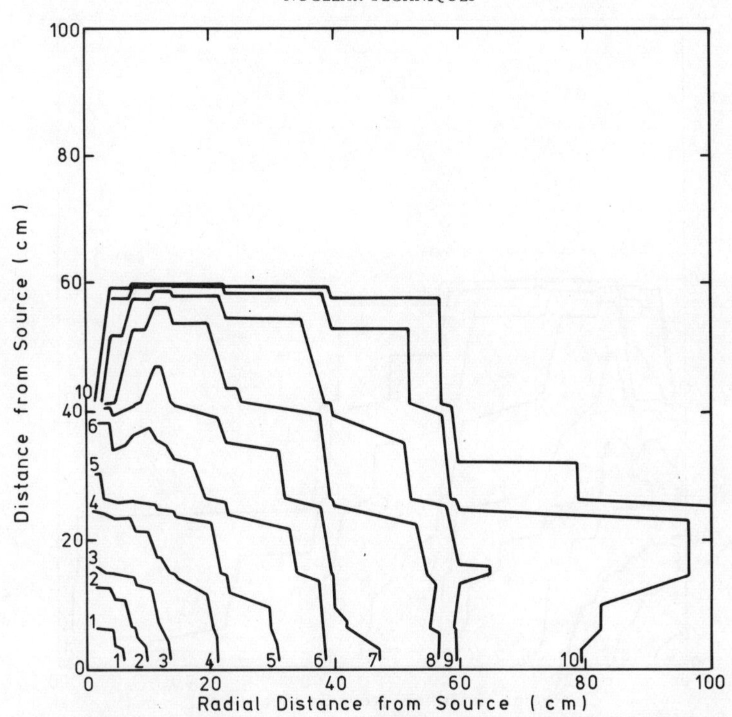

FIG. 3. Fast neutron flux distribution. Formation: silica plus 15% water. No borehole present.

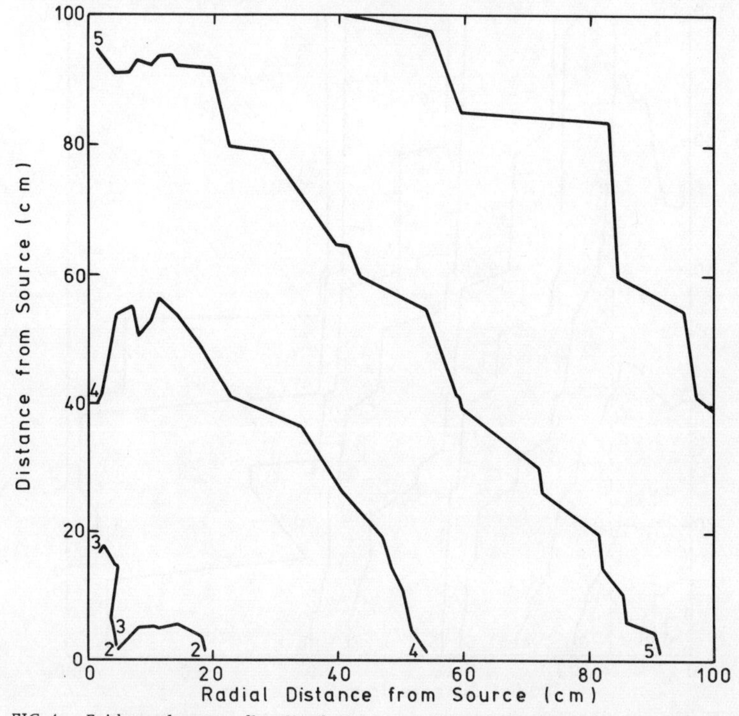

FIG. 4. Epithermal neutron flux distribution. Formation: silica. No borehole present.

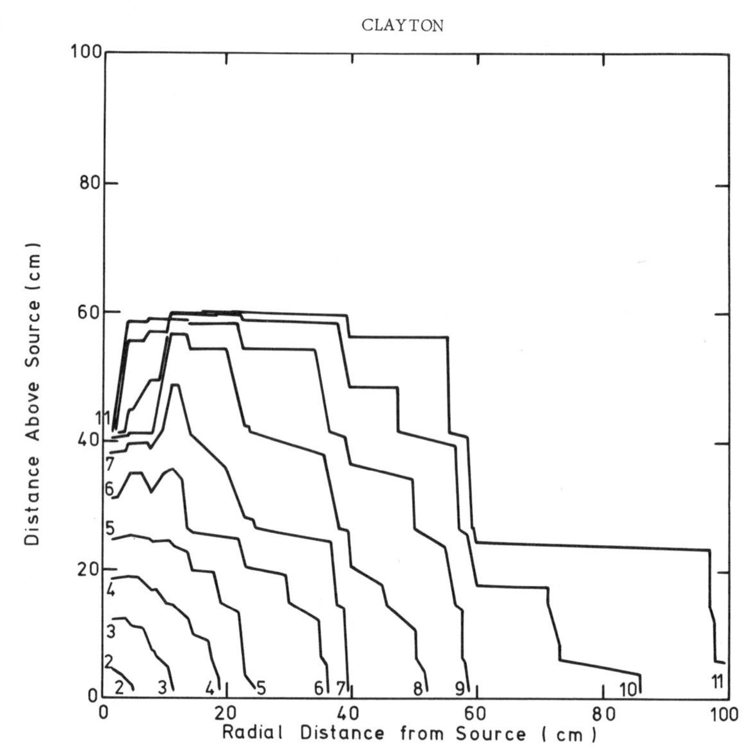

FIG. 5. Epithermal neutron flux distribution. Formation: silica plus 15% water. No borehole present.

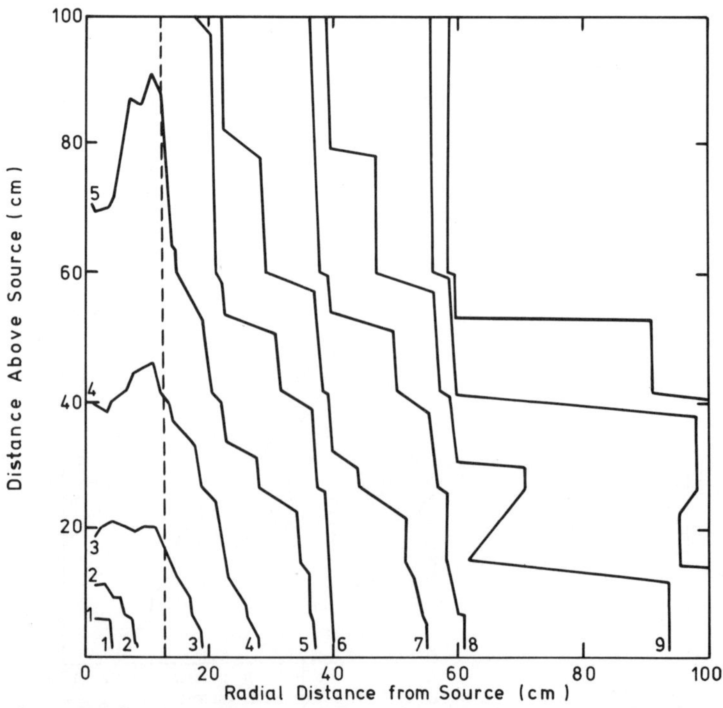

FIG. 6. Fast neutron flux distribution. Formation: silica plus 15% water. Empty borehole 13 cm in diameter.

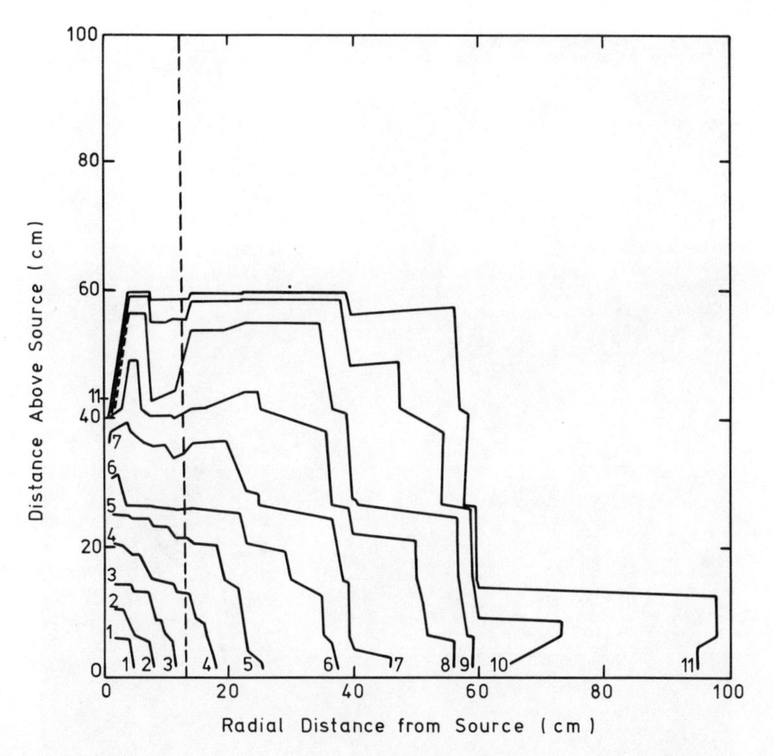

FIG. 7. Fast neutron flux distribution. Formation: silica plus 15% water. Water-filled borehole 13 cm in diameter.

The unattenuated gamma-ray fluxes at the origin due to neutron interactions in the formation have also been calculated for the fast and thermal neutron energy group and for three gamma-ray energies of 0.5, 2.0 and 8.0 MeV. The very strong effects of borehole size, water filling in the borehole and of the presence or absence of a steel liner have been demonstrated. Since the derived values are normalized to the reaction rates for a given neutron cross-section it is a straightforward procedure to apply standard renormalizing factors to account for the known cross-sections corresponding to a particular type of gamma-emitting interaction. Similarly, for the activation mode, standard time-dependent factors will renormalize to give finite irradiation and counting times.

4. THE USE OF NUCLEAR DATA IN THE DESIGN OF RADIATION INSTRUMENTS

The increasing use of nuclear techniques for mineral analysis in a wide variety of environments has emphasized the need for the presentation of nuclear data in a form which allows both a rapid assessment and a more detailed study of the most promising analytical solutions. A system of data tabulation which meets these requirements in content and form has been suggested recently by Clayton and Sanders [5].

FIG. 8. Borehole logging equipment. The trolley on the left is designed to measure the concentration of copper in "blast holes" in open-pit mines. The trolley on the right is designed to measure tin concentrations in open-pit and in underground mines.

5. SOME EXAMPLES OF THE APPLICATION AND PERFORMANCE OF ENERGY DISPERSIVE X-RAY FLUORESCENCE EQUIPMENT FOR MINERAL ANALYSIS [6]

5.1. Borehole logging equipment

Borehole logging equipment based on X-ray fluorescence techniques is finding increasing application in grade control. However, the relatively low excitation and fluorescent radiation energies associated with X-ray fluorescence analysis, especially for elements of low and medium atomic number ($Z < 40$, approx.) result in a low penetration (generally < 1 cm) into the rock and this restricts application to virtually dry and shallow boreholes. Most applications are therefore in open-pit mines and in underground mines where sufficiently dry conditions prevail. For measurement of tin and

FIG. 9. Borehole logging equipment, designed to measure the concentration of tin, arranged in the form of two back-packs.

elements of higher atomic number, operation in partially water-filled or in water-filled boreholes is possible.

By allowing percussion or rotary drilling to be used, the high cost of diamond core drilling is avoided, the cost of analysing a core is eliminated and analytical results are immediate. Because of the reduced cost, additional borehole logging can be contemplated and a more complete picture of the spatial distribution of mineralization can be obtained.

Borehole logging equipment designed to measure the concentrations of tin and of copper is shown in Fig. 8. A 'back-pack' version of the tin borehole logging unit equipment is exhibited in Fig. 9: this equipment is particularly suitable for use in underground operations in regions that are inaccessible to trolley-mounted equipment. It consists of a probe incorporating three 5-mCi americium-241 sources and a scintillation counter with balanced filters of Ag and Pd which are driven by an electric motor also mounted within the probe casing. The axial length of borehole 'sampled' at each measurement is about 5 cm. The probe is attached to a reversible scaler, mounted on a trolley or on one of the two back-packs by a cable which is normally 30 m long. However, the equipment itself is designed to operate to a depth of 300 m. All controls are on or adjacent to the scaler and the whole equipment is battery operated. The limit of detection for tin is about 0.1% (95% confidence level due to counting statistics) and this accuracy is achieved in a total measurement time of about 30 s (10 s with each filter).

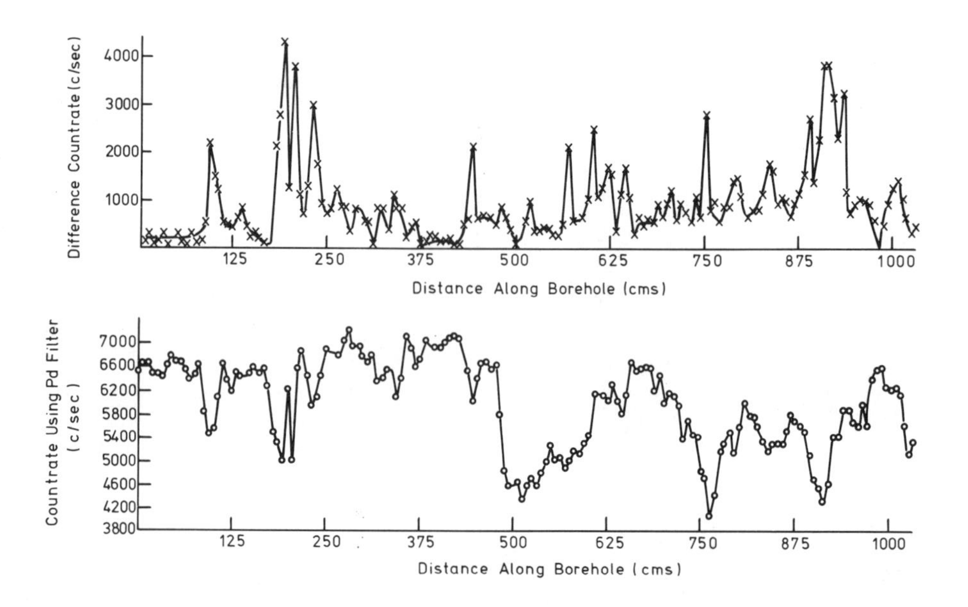

FIG. 10. Log of borehole in rock containing cassiterite.

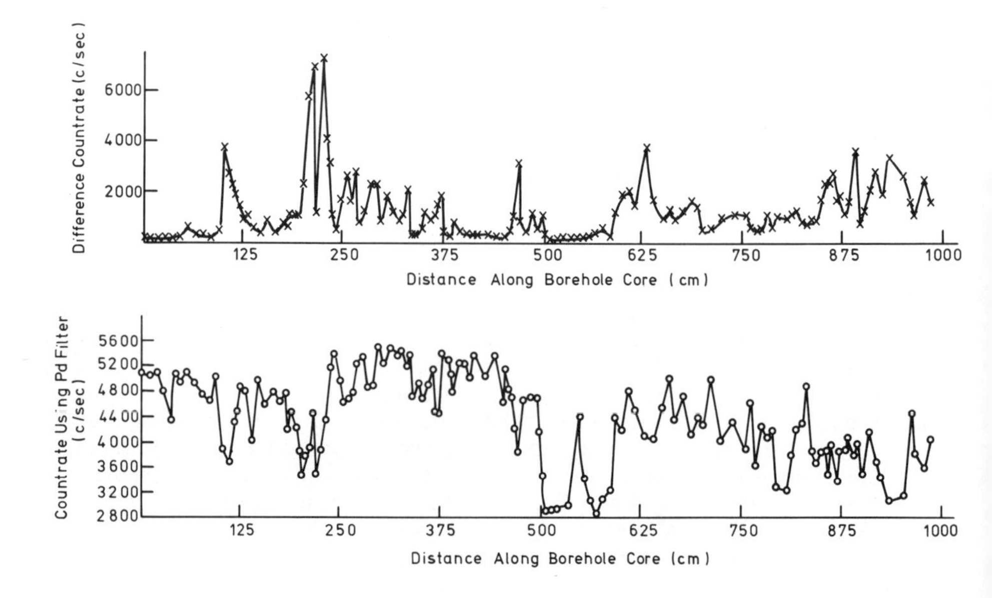

FIG. 11. Log of core removed from borehole of Fig. 9.

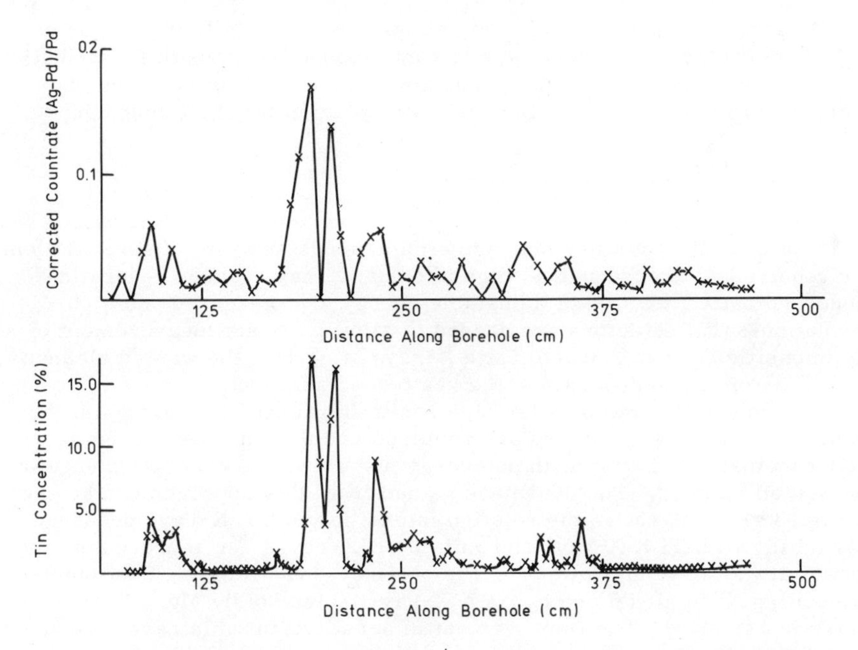

FIG. 12. Comparison between borehole log and chemical analysis of core removed from the borehole.

Figure 10 shows a typical log which gives the variations in difference count-rate along the borehole. Also shown are the variations in count-rate with the Pd filter in position: this count-rate is a measure of the scattered radiation intensity. By comparing the two count-rates over corresponding lengths of borehole it is seen that an increase in tin content (high reading in the difference count-rate) is associated with a decrease in count-rate with the Pd filter in position. However, a low count-rate from the Pd filter can also occur when the probe is in a region containing Fe but little or no tin. Such a situation occurs over a region at about 500 cm along the borehole. For comparison, a corresponding curve obtained using a portable mineral analyser to measure the concentration of tin along the length of a core from the same borehole is shown in Fig.11. Correspondence between the results obtained from the borehole log and from the core for both the differential filter measurements and for measurements using the Pd filter only is apparent, even though the tin deposit is very heterogeneous and the measurements with the probe are directed into the wall of the borehole whereas measurements using the mineral analyser on the core are virtually directed 'into the hole'.

Compensation for interference due to the presence of Fe in the rock is achieved by dividing the difference count-rate by the count-rate obtained with the Pd filter in position. The variations in tin concentration along the borehole, as indicated by this corrected reading, have been compared with the results of a chemical analysis of a sequence of 2-cm lengths of the core and this comparison is shown in Fig.12.

A reasonably good correlation is observed, especially in regions of high tin concentration. Some lack of correspondence in spatial distribution did occur because core recovery was not complete. This is a common characteristic of core analysis which is avoided in borehole logging.

5.2. Equipment for geochemical analysis

The main characteristics of equipment for geochemical analysis is that limits of detection of a few ppm and simultaneous analysis of several elements are required. In consequence, high intensity X-ray sources generating monochromatic radiation of adjustable energy are preferred and high resolution Si(Li) detectors are needed to allow separate measurement of the intensities of the characteristic X-rays of each of the wanted elements, as well as of the components of the scattered radiation.

To achieve the lowest detectable limits in geochemical analysis, the highest excitation efficiencies are required and this implies the use of monochromatic radiation with an energy just above the absorption edge of the wanted element. The advantage gained from this approach can be seen in Fig.13(a) which shows the relative intensities of S K X-rays produced by exciting with Ti K X-rays and with Mo K X-rays. An improvement in sensitivity by a factor 10 (approx.) is obtained. The improvement obtained by exciting Ni in nickel ore by Ga K X-rays instead of by Mo K X-rays is shown in Fig.13(b). The improvement in sensitivity in this case is approximately four times.

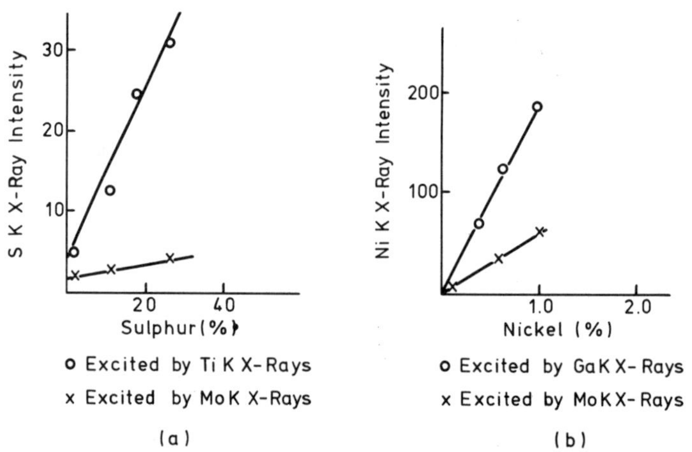

FIG.13. Improvement in sensitivity by using exciting radiation close to the absorption edge of the wanted element.

5.2.1. Single element analysis with variable matrix

The importance of selecting the energy of the exciting radiation when measuring the concentration of a single element in a sample containing much higher concentrations of several other elements, having absorption edges close together and close to the absorption edge of the wanted element, is shown by considering the analysis of Ni at concentrations of 100 ppm and 1% in the presence of Fe, Zn and Pb at concentrations of up to 15%.

Results obtained by using an X-ray tube system to generate Ga K X-rays or Mo K X-rays to excite Ni K X-rays are given in Table II and exhibited graphically in Fig.14. From this figure it is clear that excitation without correction gives severe matrix absorption for both exciting radiation energies and it is also clear that the sensitivity of the measurement using Ga K excitation is much greater than when using Mo K radiation. However, by employing corrections based on separate measurement of the Compton scattered radiation, significant improvement can be obtained by using Ga K X-rays for excitation, since the Ga K X-ray energy lies below the Zn and Pb absorption edges and above the Fe absorption edge; as does the Ni K X-rays. By using Mo K X-ray excitation, compensation for Fe is achieved, but gross over-compensation results for Zn and Pb since their absorption edges lie between the energy of the Compton scattered radiation for the Mo sources and the Ni K X-ray energy.

At a concentration of 100 ppm Ni, very good compensation is achieved, as can be seen in Table II. It should be noted that a second-order correction has been made to allow for the fraction of Fe, Zn and Pb X-rays which fall within the energy band of the detector for Ni Kα X-rays.

At concentrations of 1% Ni, under-compensation occurs for each interfering element owing mainly to the intensity of the coherent scatter radiation (which increases with increasing concentrations of Fe, Zn, Pb) lying within the Compton scatter channel ($[E_{Comp}]_{Ga}$ = 8.9 keV; $[E_{coh}]_{Ga}$ = 9.2 keV; detector resolution at FWHA = 0.18 keV).

Examination of Table II indicates the detectable limits which can be obtained by using Ga K X-rays for excitation.

The greatest improvement, which approaches a factor of 2.5, results from choosing the energy of the exciting radiation to be below the absorption edges of the interfering elements as well as being close to the energy of the Ni K absorption edge.

There is a further advantage in choosing the excitation energy below the absorption edge of the interfering elements since the total radiation intensity is then reduced and higher exciting radiation intensities can then be employed for a given total intensity incident on the detector. At the lower exciting energy there is also a significant decrease in the intensity of the scattered radiation.

5.2.2. Multi-element analysis

One of the most important advantages of using a solid-state detector in energy dispersive X-ray fluorescence analysis is that a number of elements can be analysed simultaneously.

If the wanted elements occur at similar, low concentrations, say below a few hundred ppm as in most geochemical samples, inter-element effects are small and concentrations based on measurement of the Compton scattered radiation are unnecessary. However, as the number of wanted elements in the samples increses, there is an increasing possibility of overlap between Kα and Kβ (or Lα and Lβ) radiations and allowance for this interference may be necessary. The requirement can generally be established from inspection of the emitted X-ray spectrum. A single exciting radiation energy is often sufficient for a measurement of this type.

If one or more elements are present at high concentrations (> 1%, approximately) then corrections based on measurement of the intensities

TABLE II. RESULTS OBTAINED USING Ga K X-RAYS AND Mo K X-RAYS TO ANALYSE THE CONCENTRATION OF NICKEL IN ORES CONTAINING VARYING CONCENTRATIONS OF Zn, Pb and Fe

Concentration of elements		Excitation by molybdenum K X-rays				Excitation by gallium K X-rays				
Ni as NiO (ppm)	Concentration of interfering element (%)	Uncorrected Ni Kα count[a] (100 s)	Time for 20 000 Compton-scattered Mo Kα X-rays (s)	Ni Kα count[b] corrected for 20 000 Compton-scattered counts	Error due to counting statistics (95% CL) in 1 000-s count (40 kV, 50 μA)	Uncorrected Ni Kα count[a] (100 s)	Time for 4 000 Compton-scattered Ga Kα X-rays (s)	Ni Kα count[b] corrected for 4 000 Compton-scattered counts	Error due to counting statistics (95% CL) in 1 000-s count (40 kV 50 μA)	(Error using Ga K X-ray excitation / Error using Mo K X-ray excitation) %
100	-	378 (-)	49	185	13 ppm	485 (-)	99	480	7 ppm	60
100	15 (Zn)	828 (575)	155	404	25 ppm	574 (161)	120	494	10 ppm	40
100	15 (Pb)	453 (294)	220	346	26 ppm	398 (11)	110	439	10 ppm	40
100	15 (Fe)	340 (187)	125	189	30 ppm	638 (421)	167	483	18 ppm	60
10 000	-	9760 (-)	52	5074	-	18 537 (-)	100	18 537	-	-
10 000	15 (Zn)	8391 (575)	158	12 391	-	14 331 (161)	122	17 289	-	-
10 000	15 (Pb)	6167 (294)	220	12 975	-	11 987 (11)	116	13 905	-	-
10 000	15 (Fe)	4403 (187)	130	5475	-	8851 (421)	172	14 621	-	-

[a] The counts from Zn Kα, Pb Lα and Fe Kα in the Ni Kα channel are indicated in brackets.

[b] Also corrected by subtracting 0.32% Fe Kα, 0.33% Zn Kα and 0.27% Pb Lα counts.

Note (i) Intensity of Ga K X-rays only 0.6 intensity of Mo K X-rays.
(ii) Beam currents of up to 1 mA can be used for samples with low Fe, Zn, Pb and Ni contents.

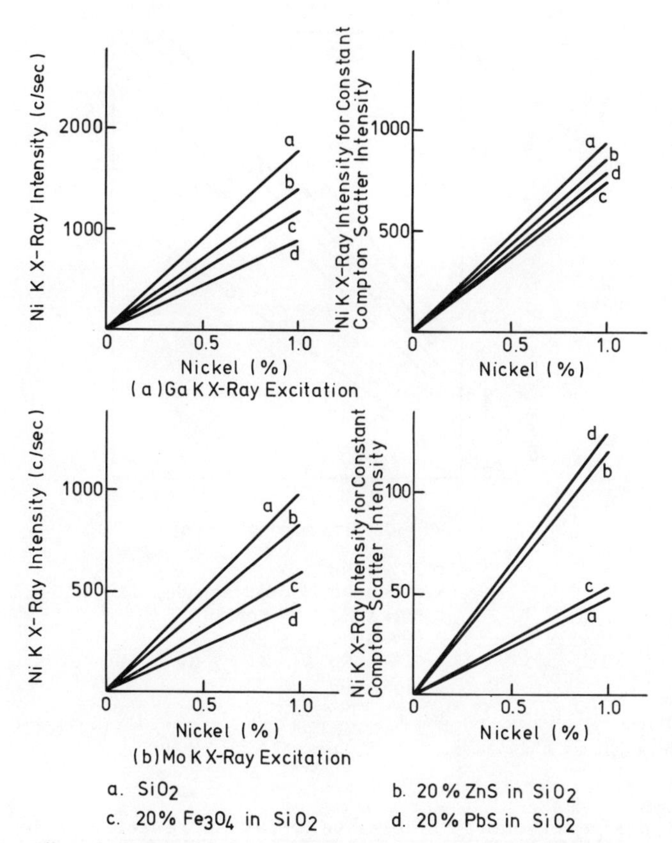

FIG. 14. Effect of compensation in a sample containing three strongly interfering elements.

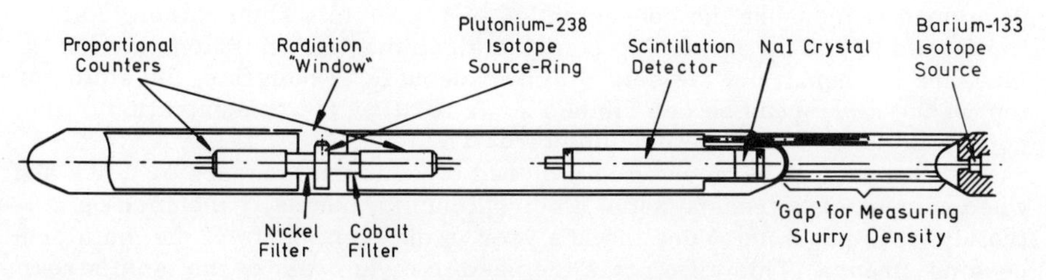

FIG. 15. General layout of the probe.

of the Compton scattered or characteristic radiations of the major interfering elements will probably be necessary. Depending on the relative atomic numbers of the major interfering and wanted elements, selection of the exciting radiation energy may or may not be critical.

As the number of interfering elements that are present at high concentration increases, greater attention must be given to the choice of the energy of the exciting radiation. More than one exciting radiation energy may be required to achieve the highest analytical accuracy and lowest limits of detection.

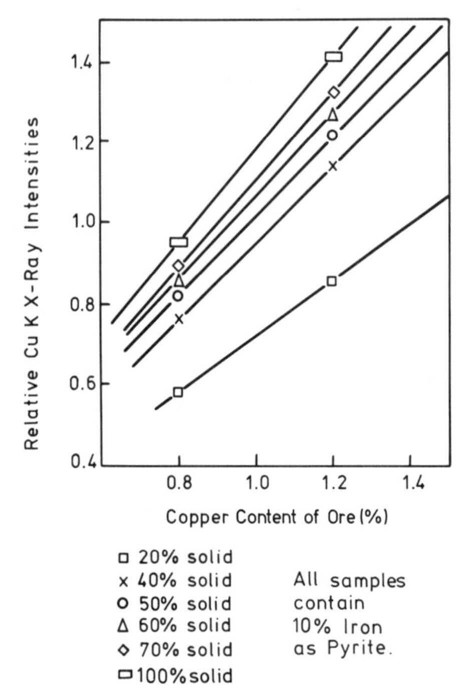

FIG. 16. Theoretical calibration curves of the intensity of Cu K X-rays as a function of the copper content of the solids for different solid contents.

5.3. An in-stream, energy dispersive X-ray fluorescence probe

An in-stream, energy dispersive X-ray fluorescence probe has been developed to measure the concentration of copper in a slurry when totally immersed [7]. The probe also contains a gamma-transmission gauge to measure the density of the slurry simultaneously. From this, the solid content of the slurry can be determined so as to allow the concentration of the copper in the ore to be determined directly.

The probe can be inserted into closed pipes, open launders or tanks and when connected to transportable electronic units, normally mounted on a trolley, it is possible to use the analyser in different parts of the main processing stream. This eliminates the need to divert part of the main stream to selected measuring positions, removes the cost of constructing sampling streams and avoids the possibility of segregation in the slurry from the use of bypass loops.

The general layout of the probe is shown in Fig.15. It includes an X-ray fluorescence section for measuring the concentration of copper and a gamma-transmission gauge for measuring the density and solids content. The probe diameter is 6.5 cm and the two sections are 45 cm and 25 cm long. The length of the connecting rods is 15 cm. The 'window' in the X-ray fluorescence section is made of polypropylene 0.75 mm thick and the entire probe, which is made of an aluminium alloy, is normally covered with 'Neoprene' before insertion in the slurry.

The X-ray fluorescence measurements are based on a double-proportional counter of cylindrical construction. It consists of a beryllium tube with a window 0.25 mm thick in the centre and a concentric anode wire with central insulator. There is a single gas filling. Each counter is surrounded by one of a pair of nickel/cobalt balanced filters. Three, 30-mCi plutonium-238 sources are mounted on a 'ring' in the central plane of the two counters.

The gamma-transmission gauge is based on a 3-mCi barium-133 source and a scintillation detector.

Figure 16 shows an experimental curve of the intensity of Cu Kα X-rays, expressed as the difference count-rate between the two proportional counters, as a function of the copper content of solids, for 60% and 70% solids content. The accuracy is independent of solids content over this range of solids. The error due to counting statistics is equivalent to ± 0.05% in the slurry at 60% solids content in material containing 0.5% Cu and 10% Fe in the solids.

In the density gauge section, the error due to counting statistics (95% CL) is approximately equivalent to ± 0.4% solids at 60% solids content.

The results indicate that this equipment has an adequate accuracy for measurement of the concentration of "heads" but is at best only marginally acceptable for measurement of tailings concentrations. For this application an immersible Si(Li) detector is recommended and this would have the advantage of enabling the concentrations of several elements to be measured simultaneously.

6. A COAL ASH MONITOR

An instrument to measure the ash content of coal based on techniques described by Rhodes et al. [8], Boyce et al. [9] and Cammack [10] has been developed in collaboration with the National Coal Board.

TABLE III. RESULTS WITH NCB/AERE ASH MONITOR FOR DIFFERENT COALS

Coal source	Ash range	Accuracy of ash measurement at 95% confidence limits
Lynemouth (lab. test)	21.9 - 34.6	± 1.64%
Monktonhall	19.5 - 35.3	± 2.0%
Bentinck (lab. test 1)	10.1 - 40.1	± 2.2%
Markham	8.1 - 32.3	± 1.27%
Seafield	8.9 - 33.7	± 3.16%
Llanharen	5.0 - 12.0	± 2.2%
Lynemouth (site test)	17.0 - 24.1	± 1.28%
Bevercotes	13.2 - 27.3	± 1.73%
Bentinck (lab. test 2)	10.0 - 40.0	± 2.0%
Easington	3.9 - 5.0	± 0.3%
Monkton	3.8 - 6.3	± 0.8%

This equipment is now installed at several collieries of the National Coal Board where it is used to control blending processes. One installation has been in operation for about three years and it is as a result of the success of this and of more recent installations that a wider use for this equipment is now being contemplated within the National Coal Board. The performance of the equipment for a number of different coals is given in Table III.

REFERENCES

[1] BOWIE, S.H.U., CLAYTON, C.G., Gamma spectrometer for sea- or lake-bottom surveying, Inst. Min. Metall., Trans., Sect. B 81 (1972) 215.

[2] MILLER, J.M., SYMONS, G.D., Radiometric traverse of the sea-bed off the Yorkshire Coast, Nature 242 (1973) 184.

[3] MILLER, J.M., ROBERTS, P.D., SYMONS, G.D., WORMALD, M.R., to be published.

[4] SANDERS, L.G., WORMALD, M.R., CLAYTON, C.G., Formation analysis by steady-state neutron interaction techniques, Soc. Prof. Well Log Analysts, Third European Symp. Trans., Paper L (1974).

[5] CLAYTON, C.G., SANDERS, L.G., "The use of nuclear data in the design of radiation instruments for mineral exploration and mining," Nuclear Data in Science and Technology (Proc. Symp. Paris, 1973) 2 IAEA, Vienna (1973) 391.

[6] CLAYTON, C.G., PACKER, T.W., The use of energy-dispersive X-ray fluorescence techniques for analysing borehole samples, Soc. Prof. Well Log Analysts, Third European Symp. Trans., Paper Q (1974).

[7] PACKER, T.W., Development of an in-stream non-dispersive X-ray fluorescence probe to determine the concentration of copper in a slurry, Harwell Industrial Report G. 97 (1973).

[8] RHODES, J.R., DAGLISH, J.C., CLAYTON, C.G., "A coal-ash monitor with low dependence on ash composition, "Radioisotope Instruments in Industry and Geophysics (Proc. Symp. Warsaw, 1965) 1, IAEA, Vienna (1966) 447.

[9] BOYCE, I.S., CLAYTON, C.G., PAGE, D., to be published.

[10] CAMMACK, P., Ash monitors for on-stream coal analysis, Conf. on Bulk Solids in Transit, Paper C59/74, London (1974).

ПЕРСПЕКТИВЫ ПРИМЕНЕНИЯ ГАММА-АКТИВАЦИОННОГО И РЕНТГЕНО-РАДИОМЕТРИЧЕСКОГО МЕТОДОВ АНАЛИЗА ПРИ ПОИСКАХ И РАЗВЕДКЕ ПОЛЕЗНЫХ ИСКОПАЕМЫХ

А.С.ШТАНЬ
Государственный комитет по использованию атомной энергии СССР,
Москва,
Союз Советских Социалистических Республик

Abstract–Аннотация

PROSPECTS FOR THE USE OF PHOTOACTIVATION AND RADIOISOTOPE-EXCITED X-RAY ANALYSIS IN MINERAL EXPLORATION.

Photoactivation analysis using bremsstrahlung from linear accelerators, betatrons and microtrons is an effective method for determining the composition of ores and ore processing products. This method has now become even more attractive with the development of high-current accelerators. Neither of the two variants of radioisotope-excited X-ray analysis (the absorption or the fluorescent method) has the same sensitivity as conventional X-ray spectroscopic analysis. However, the particular advantages offered by this method make it eminently suitable for the following applications: rapid approximate analysis of bulk materials; in-situ analysis of deposits with portable instruments; and automatic on-line monitoring of technological processes. Some instruments made in the Soviet Union are described.

ПЕРСПЕКТИВЫ ПРИМЕНЕНИЯ ГАММА-АКТИВАЦИОННОГО И РЕНТГЕНОРАДИОМЕТРИЧЕСКОГО МЕТОДОВ АНАЛИЗА ПРИ ПОИСКАХ И РАЗВЕДКЕ ПОЛЕЗНЫХ ИСКОПАЕМЫХ.

Гамма-активационный анализ с использованием тормозного излучения линейных ускорителей бетатронов и микротронов является эффективным методом для определения вещественного состава руд и продуктов их переработки. В результате разработки сильноточных ускорителей этот метод стал более перспективным. Рентгенорадиометрический анализ в своих двух модификациях (флуоресцентный и абсорбционный методы) обладает худшей чувствительностью по сравнению с классическим рентгеноспектральным анализом. Однако преимущества этого метода анализа в сочетании с достаточной чувствительностью определяют наиболее эффективные направления его применения: массовый экспресс – анализ проб; переносные приборы для анализа в естественном залегании; приборы для автоматического контроля технологических сред на потоке. Дается сводка некоторых приборов, разработанных в Советском Союзе.

I. ГАММА-АКТИВАЦИОННЫЙ АНАЛИЗ

В последние годы для определения вещественного состава пород, руд и продуктов их переработки успешно разрабатывается и применяется гамма-активационный элементный анализ [1-3].

Он заключается в определении содержания интересующих нас элементов в веществах по данным измерения наведенной активности, возникающей в результате различных фотоядерных реакций при облучении образцов γ-лучами высокой энергии.

ТАБЛИЦА I. ОСНОВНЫЕ ХАРАКТЕРИСТИКИ ЛИНЕЙНЫХ УСКОРИТЕЛЕЙ И МИКРОТРОНОВ, ПРИМЕНЯЕМЫХ В АКТИВАЦИОННОМ АНАЛИЗЕ

Название ускорителя	Тип ускорителя	Энергия ускоренных электронов (МэВ)	Средняя мощность в пучке (кВт)	Ток электронного пучка на мишенях (мкА)	Интенсивность тормозного излучения, (Р/мин · м)
ЛУЭ-8	Линейный	8,0	5–7	~700	14 000
ЛУЭ-15	Линейный	15,0	7–10	~500	55 000
МТ-20	Микротрон	20,0	0,5–1,0	~30–50	8000

Многообразие фотоядерных реакций, протекающих на большинстве ядер изотопов периодической системы элементов Менделеева, и образование при этом радиоактивных продуктов реакций позволяют применять гамма-активационный метод как метод многоэлементного анализа, обеспечивающий необходимые для практических целей экспрессность, чувствительность и точность.

Различия в пороговых энергиях и в других характеристиках фотоядерных реакций позволяют осуществлять избирательную активацию ядер изотопов различных элементов путем изменения энергии активирующего гамма-излучения.

Вместе с этим использование различий в периоде полураспада, характере и энергии излучения радиоактивных продуктов фотоядерных реакций часто позволяет полностью исключить или свести к минимуму влияние мешающих активностей, повышает селективность гамма-активационного анализа, делая однозначными его результаты.

Определяющим фактором в развитии гамма-активационных методов элементного анализа состава вещества является создание новых, пригодных для этой цели источников гамма-излучения — микротронов и линейных ускорителей, указанных в табл.1 [4,5].

Применение линейных ускорителей и микротронов позволяет проводить весьма высокочувствительный элементный анализ образцов горных пород и руд, биологических объектов, продуктов технологической переработки сырья, веществ высокой чистоты, делящихся материалов и т.п. гамма-активационным методом.

Использование электронных ускорителей для целей активационного анализа имеет целый ряд неоспоримых преимуществ:

1. Энергия, интенсивность и вид излучения (γ-кванты, нейтроны) могут быть изменены в зависимости от условий и требований анализа;

2. Потоком излучения можно управлять с помощью электромагнитных линз, устройством сканирования и т.п.;

3. Высокая проникающая способность тормозного излучения электронных ускорителей позволяет облучать образцы большого веса (до 1 кг и более), что существенно улучшает представительность анализа;

4. Ускорители имеют высокий коэффициент полезного действия;

5. Ускорители могут быть выключены в любой момент, после чего они безопасны, допускают ремонт, осмотр и т.д.

ТАБЛИЦА II. ПОРОГИ ОПРЕДЕЛЕНИЯ НЕКОТОРЫХ ЭЛЕМЕНТОВ В ОБРАЗЦАХ ГОРНЫХ ПОРОД γ-АКТИВАЦИОННЫМ МЕТОДОМ С ПРИМЕНЕНИЕМ ЛИНЕЙНЫХ УСКОРИТЕЛЕЙ И МИКРОТРОНОВ (%)*

Тип ускорителя	$10^{-1}-10^{-2}$	$10^{-2}-10^{-3}$	$10^{-3}-10^{-4}$	$10^{-4}-10^{-5}$	$10^{-5}-10^{-6}$	$10^{-6}-10^{-7}$
ЛУЭ-8	Sn	Y, Rh, Ba, W, Pt	Br, Ag, Er, Lu, Ir, Hg, Ge, Pb, U, Th	In, Cd, Au, Se, Hf, Be, D		
ЛУЭ-15	Mg, P, Fe	Al, Co, Ni, Rh, Pd, Lu	N, Cl, K, Sc, V, Mn, Cu, Zn, Ge, Nb, Ru, Cd, Ce, Nd, Dy, Er, Tm, Yb, W, Re, Pt, Pb	F, Ga, Se, Br, Rb, Y, Zr, Sb, I, Cs, Eu, Gd, Tb, Ho, Ir, Hg	Sr, Ag, In, Sn, Ba, Sm, Hf, Au, U, Th	Te, Pr, Ta
МТ-2	Mg, Al, P, Fe, Co, Ni	N, Cl, Sc, V, Zn, Ge, Nb, Ru, Rh, Pd, Ce, Gd, Dy, Er, Tm, Yb, Lu, W, Re, Ir, Pt, Pb	F, K, Mn, Cu, Ga, Se, Br, Rb, Y, Zr, Ag, Cd, I, Cs, Nd, Eu, Tb, Ho, Hf, Hg	Sr, In, Sn, Sb, Ba, Sm, Au	Te, Pr, Ta	

* Пороги определения приведены для образца весом 100 г. Время облучения образца равно времени регистрации наведенной активности и в зависимости от периода полураспада изотопа-индикатора колеблется в пределах 15с-10мин.

ТАБЛИЦА III. ХАРАКТЕРИСТИКИ НЕКОТОРЫХ РАЗРАБОТАННЫХ В СССР МЕТОДОВ ГАММА-АКТИВАЦИОННОГО АНАЛИЗА ГЕОЛОГИЧЕСКИХ ОБРАЗЦОВ И ПРОДУКТОВ ПЕРЕРАБОТКИ РУД

Исследуемые образцы	Тип ускорителя	Определяемый элемент	Порог определения (вес %)	Ссылки на литературные источники
Медные руды, вмещающие породы, продукты переработки руд	Бетатрон на 25 МэВ	Cu	10^{-2}	[1]
Свинцово-цинковые руды, вмещающие породы, продукты переработки руд	Бетатрон на 30 МэВ	Zn	10^{-2}	[1]
Полиметаллические руды, вмещающие породы, продукты переработки руд	Сильноточный бетатрон на 25 МэВ	Cu, Ba, F, Al, Mn Sb	10^{-2}– 10^{-3} 10^{-3}	[1] [14]
Титано-циркониевые руды и продукты их переработки	Сильноточный бетатрон на 25 МэВ	Ti Zr	10^{-2} 10^{-3}	[12] [12]
– " –	Бетатрон на 25 МэВ	Hf	$5 \cdot 10^{-3}$	[13]
Бериллий-вольфрам-молибденовые и медно-молибденовые руды	ЛУЭ-15	Mo	$2 \cdot 10^{-3}$	[15]
Золотосодержащие руды сложного состава	ЛУЭ-8	Au	$5 \cdot 10^{-5}$	[16]

Аналитические возможности гамма-активационного анализа с применением сильноточных электронных ускорителей приведены в табл.II [6,7].

Одной из наиболее перспективных областей применения гамма-активационного метода является анализ геологических образцов. Физические основы, а также техника и методика гамма-активационного элементного анализа проб пород и руд подробно рассмотрены в работах [8-11].

В табл.III приведены основные характеристики некоторых разработанных в СССР методов гамма-активационного элементного анализа проб горных пород, руд и продуктов их переработки. Широта охвата различных генетических типов месторождений с различной степенью сложности вещественного состава пород, руд и продуктов их переработки свидетельствует об универсальности гамма-активационного метода анализа и о возможности его применения для исследования образцов, взятых из месторождений углей, фосфоритов, циркония, меди и др.

Отсутствие верхних порогов чувствительности (некоторое исключение составляет свинец, вольфрам и другие тяжелые элементы, определение которых при содержании более 50 % затруднительно) позволяет использовать гамма-активационный анализ для исследования продуктов обогащения и металлургического передела рудного сырья. Это существенно расширяет области применения гамма-активационного анализа в горнорудной промышленности.

Электронные ускорители позволяют получать не только интенсивные потоки жесткого тормозного излучения, но и достаточно мощные нейтронные потоки. Так, ускоритель ЛУЭ-15 с бериллиевым конвертором обеспечивает выход нейтронов до $2 \cdot 10^{13}$ нейтр/с [6]. Это существенно расширяет круг аналитических задач, которые могут быть решены с помощью электронных ускорителей за счет применения нейтронно-активационного анализа в комплексе с гамма-активационным.

Из рассмотрения возможностей применения электронных ускорителей для активационного анализа можно сделать вывод о том, что на базе сильноточных линейных ускорителей электронов могут быть созданы крупные лаборатории экспрессного, высокочувствительного многоэлементного анализа образцов, обладающие высокой производительностью. Такие лаборатории целесообразно размещать в региональных центрах, расположенных поблизости от крупных промышленных предприятий, геологических экспедиций, испытывающих потребности в большом количестве анализов.

II. РЕНТГЕНОРАДИОМЕТРИЧЕСКИЙ АНАЛИЗ

Рентгенорадиометрический анализ получил за последние десять лет широкое распространение. За это время опубликовано более 500 оригинальных работ в данной области, разработано несколько десятков приборов [17].

В настоящее время рентгенорадиометрическим методом производятся экспрессные и высокоточные определения олова, тантала, ниобия, циркония, вольфрама, молибдена, бария, сурьмы, ртути и многих других элементов в рудах и продуктах их переработки.

По своей физической сущности рентгенорадиометрический метод анализа близок к классическому рентгеноспектральному анализу.

Он заключается в возбуждении атомов определяемых элементов с помощью первичного излучения от радиоактивного изотопа и последующей регистрации характеристического рентгеновского излучения возбужденных атомов с помощью специальной аппаратуры.

К настоящему времени разработаны две модификации рентгенорадиометрического анализа: флуоресцентный и абсорбционный методы.

Однако, по сравнению с рентгеноспектральным рентгенорадиометрический метод обладает пока на один-два порядка худшей чувствительностью и на данном этапе не позволяет создавать приборы для одновременного определения большого количества элементов (более пяти). Значения лучших пороговых чувствительностей рентгенорадиометрического анализа для большинства элементов ($20 \leq z \leq 92$) лежит в пределах от $n \cdot 10^{-2}$ % до $n \cdot 10^{-3}$ % (по весу). Для более легких элементов ($10 \leq z \leq 20$) пороговые чувствительности оказываются несколько хуже (от $n \cdot 10^{-1}$ до $n \cdot 10^{-2}$ %). Наконец, для отдельных элементов (хром, цирконий, ниобий, молибден, серебро, олово, йод, лантан) при измерениях в лабораторных условиях сравнительно легко достигаются чувствительности порядка $n \cdot 10^{-4}$ %. Пороги чувствительности определения ряда элементов рентгенорадиометрическим методом анализа приведены в табл. IV. К поло-

ТАБЛИЦА IV. ПОРОГИ ЧУВСТВИТЕЛЬНОСТИ РЕНТГЕНОРАДИОМЕТРИЧЕСКОГО АНАЛИЗА

Элемент	Атомный номер	Порог чувствительности (3δ, $t \le 5$ мин) в легкой матрице
Магний	12	0,1
Кремний	14	0,09
Фосфор	15	0,08
Хлор	17	0,02
Кальций	20	0,001
Титан	22	0,001
Ванадий	23	0,003
Хром	24	0,0015
Железо	26	0,004
Кобальт	27	0,001
Никель	28	0,004
Медь	29	0,003
Цинк	30	0,003
Рубидий	37	0,001
Цирконий	40	0,0005
Ниобий	41	0,0005
Молибден	42	0,00012
Рутений	44	0,03
Серебро	47	0,0005
Олово	51	0,0005
Йод	53	0,00032
Цезий	55	0,006
Барий	56	0,007
Лантан	57	0,00041
Тантал	73	0,005
Вольфрам	74	0,004
Золото	79	0,002
Ртуть	80	0,005
Свинец	82	0,004
Висмут	83	0,006
Уран	92	0,001

жительным качествам рентгенорадиометрической аппаратуры можно отнести:

1. Простота и малые габариты;
2. Невысокая стоимость;
3. Транспортабельность;
4. Большие возможности для автоматизации технологических процессов;
5. Возможность определения легких элементов (легче марганца) благодаря бескристальной системе регистрации мягкого характеристического излучения этих элементов;
6. Идеальная стабильность первичного источника излучения позволяет снизить ошибки анализа при объективном способе наблюдений;
7. Использование монохроматичных источников возбуждающего излучения позволяет уменьшить зависимость результатов анализа от вещественного состава проб и повысить точность определений.

Указанные преимущества в сочетании с достаточной для многих практических задач чувствительностью метода обуславливают широкие возможности его использования.

Они же определяют и наиболее эффективные направления применения метода:

1. Массовый экспресс-анализ технологических или геологических проб·
2. Переносные приборы для анализа в естественном залегании;
3. Приборы для автоматического контроля технологических сред на потоке.

В Советском Союзе достаточно эффективно развиваются все указанные направления. Сводка некоторых приборов для рентгенорадиометрического анализа, разработанных в СССР, приведена в табл.V.

Исторически и в Советском Союзе,и в других странах еще в начале шестидесятых годов появились приборы для экспресс-анализа порошковых проб (в основном для анализа минерального сырья). Сюда относятся, например, приборы "Минерал-2" и "Минерал-3" [17]. Постепенно эти приборы совершенствовались, оснащались специальными методиками и приспособлениями для устранения влияния вещественного состава наполнителя (эффекта матрицы) и проведения более точных и достоверных анализов, переводились на более совершенную элементную базу (микромодули, интегральные схемы, полупроводниковые детекторы, специальные источники излучения), оснащались блоками для обработки результатов измерений и устройствами автоматической смены проб. Так, в приборе АЖР-1 [18] введен второй канал для компенсации эффекта матрицы по рассеянию бета-излучения. Прибор ФРАД-1 [19] обладает более совершенной схемой компенсации эффекта матрицы и имеет нормированные метрологические характеристики.

Важным направлением работ является переход в область анализа более легких элементов, где начинает сказываться поглощение флуоресцентного излучения воздухом. В этих случаях приходится помещать измеряемые пробы в вакуумную камеру, как это делается в приборе "Маяк-В" [20].

Одной из областей применения рентгенорадиометрического анализа, где последний оказывается почти вне конкуренции за счет простоты и портативности аппаратуры, является анализ минерального сырья в естественном залегании. Разработанные для этой цели переносные приборы РПС4-01 [21] и ФАМ-3-01 [22] с успехом используются при поисках и для оценки запасов полезных ископаемых.

Значительные усилия были сосредоточены на разработке рентгенорадиометрических приборов для контроля конкретных технологических процессов. К таким приборам относятся абсорбционные концентратомеры тантала в порошковых пробах КТ-1 и ниобия в растворах КТН-3 [23], а также флуоресцентный концентратомер тантала и ниобия КТН-2 [24].

Одним из наиболее перспективных направлений работ является создание автоматических анализаторов на потоке, которые экономически наиболее эффективны, позволяют оперативно влиять на ход технологического процесса без участия оператора и способствуют быстрейшему проведению комплексной автоматизации производства. Пока подобные анализаторы сравнительно просто удается создавать только на потоке растворов. Как правило, они работают на абсорбционном принципе [23,25] и обладают весьма высокой селективностью к определяющему элементу.

ТАБЛИЦА V. СВОДКА НЕКОТОРЫХ ПРИБОРОВ ДЛЯ РЕНТГЕНОРАДИОМЕТРИЧЕСКОГО АНАЛИЗА, РАЗРАБОТАННЫХ В СССР

Наименование прибора	Назначение	Принцип действия	Основные параметры
АЖР-1	Экспресс-анализ элементов группы железа в порошковых пробах	Флуоресцентный с компенсацией эфф. матрицы по β-рассеянию	Пределы – 1 ÷ 70% Точность – 0,5% Fe Время измерения – 1 ÷ 2 мин
КТН-1	Контроль концентрации тантала и ниобия в растворах на потоке	Абсорбционный по скачку поглощения	Пределы – 0,1 ÷ 20 г/л Точность – 0,05 г/л+1%
ФРАД-1	Экспресс-анализ элементов $20 \leq z \leq 92$ в порошковых пробах	Флуоресцентный с автостабилизацией, 2 канала	Пределы – 0,1 ÷ 100% Точность – 0,1% абс + 0,005 Время измерения – 3 мин
КТН-1	Экспресс-анализ тантала в порошковых пробах	Абсорбционный по скачку поглощения	Пределы – 0,1 ÷ 100% Точность – 0,01 Ta + 1% Время измерения – 10 мин
КТН-2	Экспресс-анализ тантала и ниобия в порошковых пробах	Флуоресцентный (метод спектральных отношений)	Пределы – 0,1 ÷ 100% Точность – 0,01 абс + 1% Время измерения – 10 мин
КТН-3	Экспресс-анализ тантала и ниобия в растворах	Абсорбционный по скачку поглощения	Пределы – 0,1 ÷ 100% Точность – 0,01 г/л + 1% Время измерения – 10 мин
ФАМ-3-01	Опробование молибденовых руд в естественном залегании	Флуоресцентный с вычетом фона	Пределы – 0,02 ÷ 4% Точность – 0,025% абс + 0,05 Время измерения – 3 мин
МАЯК-Р	Экспресс-анализ элементов $24 \leq z \leq 92$ в порошковых пробах	Флуоресцентный с автостабилизацией, 4 канала	Порог – 10^{-2}% Точность – 1 ÷ 7% Время измерения – 2 ÷ 15 мин
МАЯК-В	Экспресс-анализ элементов $11 \leq z \leq 24$	Флуоресцентный с вакуумной камерой	Порог – 10^{-1}% Точность – 5 ÷ 10% Время измерения – 2 ÷ 10 мин
РПС 4-01	Экспресс-анализ в естественном залегании и в порошковых пробах элементов $22 \leq z \leq 92$	Флуоресцентный с дифференциальными фильтрами	Порог – 10^{-2}% Точность – 5% Время измерения – 1 мин

Широкое внедрение автоматических анализаторов в промышленность потребует решения ряда научных и технических задач, таких как:

1. Разработка специальных пробоотборных и пробоподготовительных устройств.

2. Поиск новых инструментальных приемов снижения влияния состава наполнителя на результаты анализа.

3. Расширение области применимости метода в сторону более легких элементов.

4. Повышение чувствительности и точности аппаратуры.

5. Поиск путей создания приборов для одновременного определения концентрации нескольких элементов с использованием для автоматической обработки результатов анализа ЭВМ.

Решение указанных задач в значительной степени зависит от разработки более совершенных источников низкоэнергетического рентгеновского и альфа-излучения, необходимых для анализа легких элементов. Громадное значение имеет совершенствование полупроводниковых детекторов с разрешением на уровне 150-200 эВ, а также создание ППД, работающих при комнатных температурах (CdTe, AsGa и др.). Именно на базе подобных детекторов можно будет создавать надежные автоматические анализаторы для одновременного определения нескольких элементов на потоке.

ЛИТЕРАТУРА

[1] СУЛИН, В.В., В сб. Ядерно-физические Методы Анализа Вещества, Атомиздат, М., 1971, стр 206.

[2] ШТАНЬ, А.С. и др., 4th Int. Conf. Peaceful Uses At. Energy (Proc. Conf. Geneva, 1971) 14, UN, New-York and IAEA, Vienna (1972) 183.

[3] БУРМИСТЕНКО, Ю.Н., "Состояние и перспективы развития аппаратуры и методов активационного анализа вещества с использованием излучений электронных ускорителей", Тезисы докладов Всесоюзного научно-технического семинара "Применение ускорителей в элементном анализе различных веществ", ВДНХ, М., 1973, стр.8.

[4] КОМАР, Е.Г., 4th Int. Conf. Peaceful Uses At. Energy (Proc. Conf. Geneva, 1971) 14, UN, New-York and IAEA, Vienna (1972) 321.

[5] Микротрон, Сб.статей, Изд-во АН СССР, 1965.

[6] ШТАНЬ, А.С., Атомная Энергия 33 4 (1972) 858.

[7] БУРМИСТЕНКО, Ю.Н., РЫВКИН, Б.Н., ФЕОКТИСТОВ, Ю.В., "Изучение характеристик продуктов активации, полученных на электронном ускорителе ЛУЭ-15", Тезисы докладов Всесоюзного научно-технического семинара "Применение ускорителей в элементном анализе различных веществ", ВДНХ, М., 1973, стр.37.

[8] СУЛИН, В.В., В сб. Ядерная Геофизика, Гостоптехиздат, М., 1963, стр.167.

[9] Фотоядерные методы анализа вещественного состава горных пород и руд с применением бетатронов, Тематический выпуск бюллетеня научно-технической информации, Серия: Региональная, разведочная и промысловая геофизика, №6, Изд. ОНТИ ВИЭМС, 1967.

[10] Труды ВНИИЯГГ, вып.1., Ядерная Геофизика, М., "Недра", 1968.

[11] NIEWODNICZANSKI, I., et al., Nucleonica, 12 (1967) 1053.

[12] БЕРЗИН, А.К., ВИТОЖЕНЦ, Г.Ч., МЕЩЕРЯКОВ, Р.П., СУЛИН, В.В., ШОРНИКОВ, С.И., ЯКОВЛЕВ, Б.М., "Применение сильноточного бетатрона "Луч" (Б-25) для высокочувствительного γ-активационного анализа проб пород и руд на медь, цирконий и титан", В сб. Недеструктивные Активационные Методы Анализа Некоторых Видов Минерального Сырья и Применяемая Аппаратура, ОНТИ ВНИИЯГГ, М., 1972, стр.33.

[13] МАРТЫНОВ, Ю.Т., СУЛИН, В.В., ШОРНИКОВ, С.И., "Некоторые результаты разработки и опробования методики γ-активационного определения содержания гафния в пробах титано-циркониевых руд и продуктов их переработки", Там же, стр.23.

[14] ВИТОЖЕНЦ, Г.Ч., СУЛИН, В.В., ХРЫНИН, Б.А., "О применимости γ-активационного метода для высокочувствительного и точного определения содержания сурьмы в пробах полиметаллических руд и вмещающих горных пород", Там же, стр.44.

[15] БУРМИСТЕНКО, Ю.Н., ВИТОЖЕНЦ, Г.Ч., СУЛИН, В.В., ФЕОКТИСТОВ, Ю.В., ХРЫНИН, Б.А., "Исследование практических возможностей применения γ-активационного метода для определения содержания молибдена в пробах горных пород и руд", Там же, стр.64.

[16] БУРМИСТЕНКО, Ю.Н., ГАМБАРЯН, Р.Г., ИВАНОВ, И.Н., ФЕОКТИСТОВ, Ю.В., ШТАНЬ, А.С., "Разработка экспрессного, высокопроизводительного метода определения золота в рудных порошковых пробах на основе применения реакции неупругого рассеяния гамма-квантов пучка тормозного излучения сильноточного линейного ускорителя электронов", Тезисы докладов Всесоюзного научно-технического семинара "Применение ускорителей в элементном анализе различных веществ", ВДНХ, М., 1971, стр.44.

[17] ЯКУБОВИЧ, А.Л., ЗАЙЦЕВ, Е.И., ПРЖИЯЛГОВСКИЙ, С.М., Ядерно-физические Методы Анализа Минерального Сырья, Атомиздат, М., 1969.

[18] МАМИКОНЯН, С.В., ВАРВАРИЦА, В.П. и др., "Промышленный образец рентгенорадиометрического анализатора концентрации элементов группы железа АЖР-1", В сб. Радиационная Техника, Атомиздат, 7 (1972) 157.

[19] МАМИКОНЯН, С.В., МЕЛЬТЦЕР, Л.В. и др., "Широкодиапазонный двухканальный рентгенорадиометрический анализатор ФРАД-1", Изотопы в СССР, 39 (1974) 12.

[20] ЯКУБОВИЧ, А.Л., БАБИЧЕНКО, С.И. и др., "Унифицированная аппаратура для анализа минерального сырья комплексом ядернофизических методов", В сб. Ядернофизические Методы Анализа Вещества, Атомиздат М., 1971, стр.384.

[21] ЯКУБОВИЧ А.Л., БАБИЧЕНКО, С.И. и др., "Рентгенорадиометричесий анализатор для ускоренной оценки элементного состава руд и продуктов их переработки без отбора проб. В сб. Ядернофизические Методы Анализа Вещества, Атомиздат, М., 1971, стр.388.

[22] АШИТОК, В.И., БАЛДИН, С.А. и др., "Применение рентгенорадиометрического анализатора для опробования руд Хибинских апатитовых месторождений", В сб. Ядерное Приборостроение, Атомиздат, М., 21 (1973) 40.

[23] КОХОВ, Е.Д., ЛАВРЕНТЬЕВ, Ю.Ж., МАМИКОНЯН, С.В., "Абсорбционные рентгенорадиометрические приборы технологического контроля состава вещества", Изотопы в СССР, 38 (1974) 10.

[24] КОЛЕСНИКОВ, Б.Е., КОХОВ, Е.Д., "Концентратомер тантала и ниобия КТН-2", Атомная Энергия 37 1 (1974) 56.

[25] ИСАЕВ, В.С., КОКОЛЕВСКИЙ, В.И. и др., "Концентратомер тантала и ниобия КТН-1". В сб. Радиационная Техника, Атомиздат, 8 (1972) 194.

PRACTICAL EXPERIENCES OF VARIOUS NUCLEAR TECHNIQUES SUPPORTING MINERAL PROSPECTING IN GREENLAND

L. LØVBORG, H. KUNZENDORF,
E. MOSE CHRISTIANSEN
Danish AEC Research Establishment Risø,
Roskilde, Denmark

Abstract

PRACTICAL EXPERIENCES OF VARIOUS NUCLEAR TECHNIQUES SUPPORTING MINERAL PROSPECTING IN GREENLAND.

The Electronics Department at Risø assists the Geological Survey of Greenland in the prospecting for uranium and other metals. The techniques used in the field or in the laboratory or in both include gamma-ray spectrometry, delayed-neutron counting, radioisotope-excited X-ray fluorescence and gamma-activation analysis of beryllium. The practical experiences obtained with these techniques are described in the form of a summary of material that has been published during the years 1969-1974. Future activities will mainly be based on airborne gamma-ray spectrometry and automatic sample analysis by delayed-neutron counting and X-ray spectrometry.

1. INTRODUCTION

In 1956 the Kvanefjeld uranium deposit [1] in south Greenland was discovered. It became a natural task for the Electronics Department to assist the Geological Survey of Greenland in the radiometric evaluation of the ore reserves. Surface mapping by portable gamma-ray spectrometers was introduced in 1966 [2, 3], and a gamma-spectrometric scanning of 3 500 metres of drill core was carried out in 1969-70 [4]. Along with these developments the co-operation between the Department and the Survey was extended to include prospecting for beryllium, zirconium and niobium in south Greenland. Beryllometers were constructed [5, 6], and field work with portable X-ray fluorescence analysers was accomplished [7, 8].

Since 1971 the applications of nuclear techniques have been concentrated on primary mineral prospecting in central east Greenland. This part of the country differs considerably from the south-western part with respect to geology, climate, and accessibility. The chief purpose of the activities in east Greenland is uranium exploration. Field operations are based on airborne gamma-ray spectrometry, radiometric follow-up work, detailed geological mapping, and sampling of soils, stream sediments and river waters. Since the field seasons are short, and the quartering primitive, the contents of uranium and other metals in the samples are not being assayed in the field. Instead, much more emphasis is being attached to laboratory analysis than before.

As regards analysis for uranium, the most significant development is a delayed-neutron counting facility which has been installed at one of the research reactors at Risø. The availability of this powerful tool has made it possible to begin geochemical prospecting for uranium on a large scale.

In the prospecting for ore metals, particularly copper, zinc, lead, and molybdenum, energy-dispersive X-ray spectrometry has proved very promising [9, 10].

Here, an attempt is made to summarize what we have learned regarding the practical applicability of the various nuclear techniques used by us in the field and in the laboratory. Though the conditions for mineral prospecting in Greenland are rather special, owing to the arctic climate of the country and its remote location from domestic Denmark, Greenland is similar to many countries under development regarding the logistic difficulties that arise from a scarcity of roads and service facilities.

2. RADIOMETRIC TECHNIQUES

Since uranium exploration is the principal basis of our co-operation with the Geological Survey, we have spent much time in developing gamma-spectrometric prospecting and evaluation methods. Besides, a considerable part of our research activities is devoted to problems in connection with the interpretation of ground and airborne radiometric surveys [11, 12].

2.1. Laboratory gamma-ray spectrometry

The Kvanefjeld uranium deposit (reasonably assured reserves, 5 800 metric tons of uranium; average ore grade, 310 ppm U) is unique in the sense that the Th/U ratio of the ore body varies from less than one to more than five over short distances. For this reason it was considered inconvenient to evaluate the uranium resources on the basis of drill-hole logging. Since the drill cores were of a very good quality, it was decided to build a special gamma-ray spectrometer by means of which 1-m-long core sections could be scanned (Fig. 1).

This project turned out to be a technical success [4, 13]. Core scanning was found to be faster and just as accurate as chemical or gamma-spectrometric whole-rock assays of crushed core, but compared with drill-hole logging it was ten times slower. Several of the holes were logged, and the log-diagrams were compared with the U-Th contents in the cores. In spite of the very heterogeneous nature of the ore deposit, the correlation coefficient of the two sets of data was greater than 0.7.

Laboratory gamma-ray spectrometry of geological samples is now carried out to a limited extent. As the materials handled in our laboratory have widely different radioelement contents we have not considered it practical to aim at low-level gamma-ray counting.

Our general-purpose gamma-ray spectrometer is operating automatically [13]. It is based on the use of a single 5 in × 5 in NaI(Tl) detector equipped with a 3-in-thick NaI light guide to suppress the ^{40}K gamma radiation from the photomultiplier. A 15-cm-thick cylinder made of antimony-lead serves as a background shield. The sample materials, usually coarsely crushed rock specimens, are poured into 150-ml plastic canisters, the lids of which are sealed with glue. We are aware of the fact that plastic is permeable to radon, but we have never noticed any radon loss from our samples. The uranium and thorium standards used contain 0.1% of the respective radioelement in a matrix of dunite (NBL Nos 74 and 80). Our potassium standard

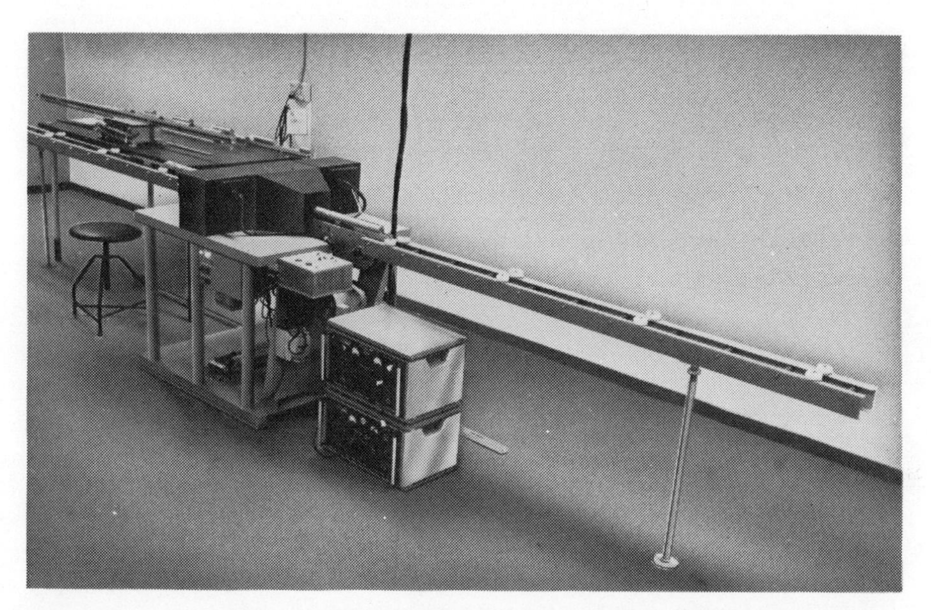

FIG.1. Instrument for gamma-spectrometric scanning of drill-core sections. The core is viewed by two opposing 2 in dia. × 3 in NaI(Tl) crystals housed in a lead shield. A continuous scan of a 1-m-long core section typically lasts 20 min.

is a sample of pure KCl. Though the abundance of potassium in KCl is 52.45%, we have found it necessary to compensate for the low gamma-ray absorption of this material by ascribing to it a content of 56.8% potassium.

We have tried to analyse samples on the basis of multichannel spectrum recording followed by a least-squares data reduction method, but found that simple, three-channel gamma-ray counting worked just as well. Energy limits, background count-rates and sensitivities for the counting channels routinely used are given in Table I. A spectrometer run of typically 40 samples lasts as long as 4 days; each sample is counted three times and each count lasts 40 min. For common sample materials the practical detection limits are of the order of 0.5 ppm U, 1 ppm Th and 0.1% K.

2.2. Field gamma-ray spectrometry

In a previous paper [14], one of the authors advocated the use of portable gamma-ray spectrometers for primary uranium prospecting in difficult terrains. With regard to uranium exploration in Greenland, instruments like these have only been used in the traditional way, i.e. for detailed radiometric mapping of small areas.

In our first attempt [2] to map the surface distribution of uranium and thorium at Kvanefjeld we learned that a portable gamma-ray spectrometer is of little value without a suitable calibration facility. Our next attempt was much more successful [3]. This time we used a 3 in × 3 in NaI(Tl) detector which had been calibrated by means of two experimental concrete cylinders containing uranium and thorium in known amounts. To

TABLE I. BACKGROUND COUNT-RATES AND SENSITIVITIES OF LABORATORY GAMMA-RAY SPECTROMETER FOR ANALYSIS OF Th-U-K IN ROCK SAMPLES

	Energy interval (MeV)		
	1.30 - 1.60	1.65 - 1.95	2.45 - 2.75
Background (counts/min)	35	20	10
Sensitivities (counts/min per mg)			
Thorium	2.5	1.4	2.9
Uranium	12	12	0.3
Potassium	0.011	0	0

FIG.2. Array of four concrete platforms for calibration of radiometric field instruments, in particular portable gamma-ray spectrometers. Each platform is 3 m in diameter and 0.5 m thick and contains evenly dispersed radioactive minerals.

FIG.3. Gamma-spectrometric field assay of granite outcrop in east Greenland. The gamma-ray detector (a 3 in dia. × 3 in NaI(Tl) crystal) is worn on the back, whereas the four-channel pulse-height analyser is mounted on the front of the prospector.

ensure that the characteristic, sharp variations in the radioactivity of the Kvanefjeld rocks could be resolved, the detector was housed in a stainless-steel pot lined with 5 cm of lead.

Since then, field-spectrometric mapping of selected areas in east Greenland has been carried out with the purpose of making these areas applicable as reference targets for our airborne gamma-ray spectrometer. The problem of calibrating the field instrumentation has been finally solved through the construction at Risø of an array of four calibration platforms, 3 m in diameter and 0.5 m thick [11], see Fig. 2. Our measurements in east Greenland have shown that a 3 in × 3 in NaI(Tl) detector, mounted 1 m above the ground on a tripod or worn on the back (Fig. 3), is adequate for field assay of outcrops with radioelement contents similar to those found in a typical granite. For the time being we are planning to make use of portable gamma-ray spectrometers to supplement the simple survey meters used until now in the exploration of radioactive anomalies disclosed by the airborne measurement system.

2.3. Airborne gamma-ray spectrometry

The first aeroradiometric survey of central east Greenland was carried out by means of a Dornier-28 aircraft equipped with two 6 in dia. × 4 in NaI(Tl) detectors and an analogue-recording, four-channel pulse-height analyser. A radar altimeter monitored the terrain clearance, which we attempted to maintain at 50 m. After the flights, the strip-chart records were digitised using an optical curve follower at the hybrid computer installation at Risø by which compact, radiometric profiles could be produced on a digital plotter [15]. Attempts were made to correct the data for variations in the survey height, but the commonly used height-correction formulae were found to be inadequate, one of the reasons being that most flights were made in mountainous terrain. Approximate stripping ratios for the airborne spectrometer were determined by means of our concrete calibration platforms. Apart from a single case, in which the stripped count-rates allowed us to conclude that a radioactive anomaly was produced by thorium minerals in the ground, the window count-rates were too low to permit them to be stripped.

An improved, airborne gamma-ray spectrometer, designed and built by the Electronics Department [16], was installed in a Britten-Norman Islander aircraft and put into operation in 1973. The detector unit of this instrument contains six thermostat-controlled 6-in dia. × 4 in crystals. The pulses from the detector are processed by means of standard NIM modules and a single-crate CAMAC system is used for control and readout purposes. After each counting period, typically lasting one second, the contents of the four scalers and the signal from the radar altimeter are stored. At the same time a new counting period is started during which the stored information is punched in binary code on paper tape. Timing information is punched every

FIG.4. Twin-engine aircraft (Britten-Norman Islander) parked on natural air-strip near Moskusokse Fjord, east Greenland. The aircraft is equipped with a four-channel, digital gamma-ray spectrometer (crystal volume 11 120 cm^3), a proton magnetometer and a 35-mm camera.

32nd counting period. To facilitate the recovery of the flight lines, a 35-mm camera simultaneously takes pictures of a time indicator and the ground beneath the aircraft.

This complex measurement system has been flown at a terrain clearance of 100 m under rather severe conditions. The aircraft was operated from a natural air-strip at 73° 40'N where a camp of tents had been established (Fig. 4). Since the aircraft was also used for transport purposes, the spectrometer units had to be dismounted and re-installed a number of times. Cable leads and connector pins unavoidably suffered from this handling which meant that the survey program could not be accomplished without skilled technicians at hand.

The use of punched paper tape was practical because we could control the data on the spot by direct visual inspection. On the other hand, special programs had to be developed before the tapes could be processed at the Borroughs B-6700 computer available to us at Risø. Spectrometric data recorded during more than 200 flying hours are now stored on magnetic tape from which they can be retrieved as tables and graphs.

3. DELAYED-NEUTRON COUNTING

Our delayed-neutron counting facility is mostly used for pure uranium assays, but combined uranium-thorium analysis of crushed rock and mineral concentrates is also carried out. The sample materials are poured into 6-ml polyethylene ampoules which are sealed by means of ultrasonic welding. Reference samples are prepared from standard solutions of U_3O_8 and ThO_2. In the assay of solid material, the reference samples contain 0.6 ml of solution absorbed in silica gel.

The irradiation and counting facility is located in the basement of the 5-MW research reactor DR 2, see Figs 5 and 6. At the start of a measurement, the sample is placed in a 20-cm long plastic rabbit. The rabbit is transferred pneumatically to a position at the edge of the reactor core where the thermal flux, the fast flux, and the gamma-ray exposure rate are 1×10^{13} n/cm^2·s, 2×10^{12} n/cm^2·s and 150 MR/h, respectively. After one minute, the rabbit automatically returns to the handling cell in which it is opened by means of a pair of manipulators. The radioactive specimen is dropped into a metal tube which ends in the middle of a water-filled tank containing a circle of nine BF_3 counter tubes. The counting of the pulses from these begins 20 or 24 seconds after the rabbit has left the reactor. One minute later, the irradiation time, the delay time and the number of delayed fission neutrons detected are printed and punched on a teletype printer, after which the sample is lifted back to the handling cell.

In combined uranium-thorium analysis, the samples must also be counted after having been irradiated in a cadmium-lined rabbit. The cadmium-ratio for samples containing little thorium is approximately 22. Duplicate measurements are always made. If the results of these are not consistent, the sample in question is measured once more. Background determinations are made at frequent intervals. In the assay of liquid samples a blank is required, whereas the background in assaying terrestrial material is measured without using a blank. About 60 background counts are typically registered, while roughly 800 and 7 counts are produced per microgramme of uranium and thorium, respectively.

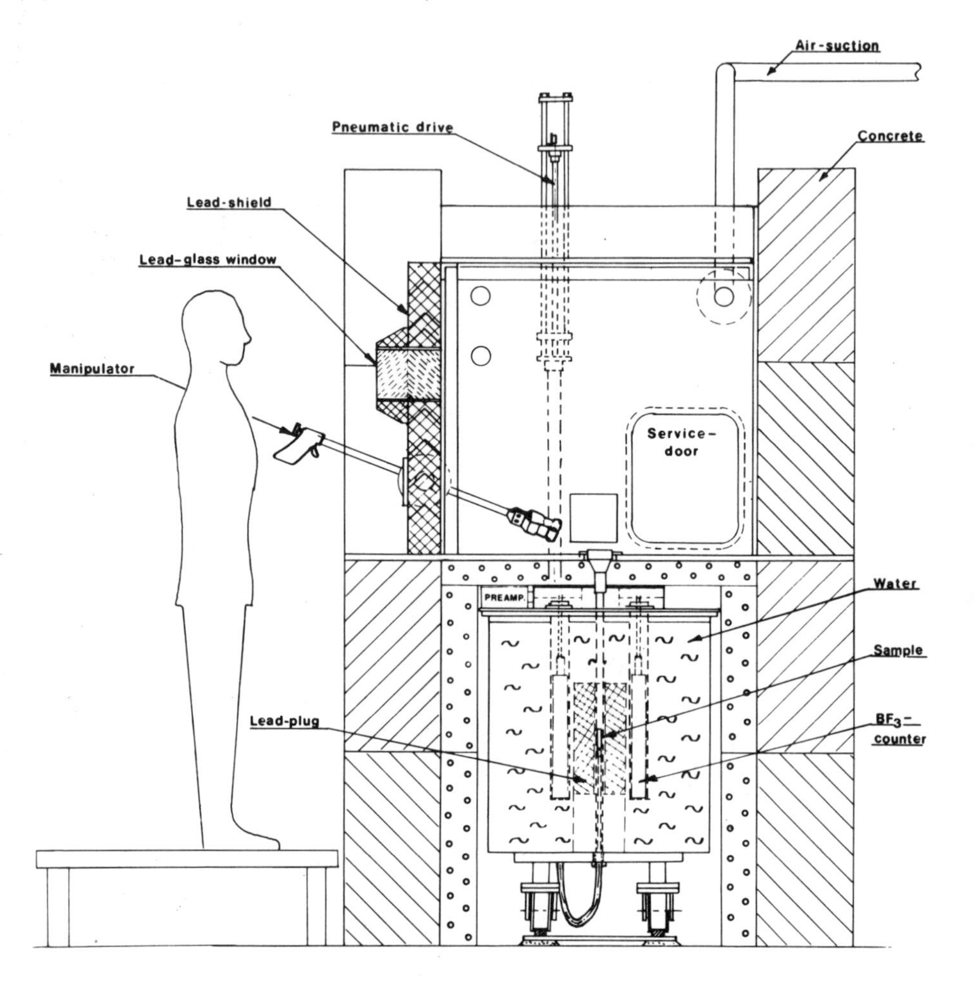

FIG.5. Schematic view of delayed-neutron counting facility at the Danish research reactor DR 2 for determination of uranium and thorium in geological samples.

The only problem we have had with the counting facility was the counting of noise pulses induced by the electrical machinery within the reactor containment. This problem was partly overcome by the introduction of inductance in the connection leads between the counting equipment and the neutron detectors. Spurious counts do occur now and then. These are discarded by using the Dixon criterion of rejecting extreme observations [17] in the computer program, by means of which the tapes from the teletype printer are processed.

About 40 uranium assays or 12-15 uranium-thorium assays can be made in one day. Analytical results are stated according to the scheme proposed by Currie [18]. For example, in assaying uranium the background count-rate determines a critical level, which the net count-rate from the sample

FIG.6. Transfer of sample to be irradiated in the DR-2 reactor. The irradiation time, the delay time and the counting of delayed neutrons are controlled by means of the instrument panel on the right-hand side. Timing information and counting results are punched by the teletype printer on the left.

must exceed, before uranium can be said to have been detected in the sample. For uranium the detection limit is of the order of a few pp 10^9, for thorium it generally amounts to several ppm. Whereas maximum concentrations are always estimated, analytical determination values are only reported if their standard deviations are less than 25%.

4. RADIOISOTOPE-EXCITED X-RAY FLUORESCENCE

In our studies of the applicability of radioisotope-excited X-ray fluorescence for mineral-prospecting purposes, we have made use of portable instruments, operating on the basis of X-ray filters [7, 8, 19], and semiconductor-detector X-ray spectrometers [9, 20]. The results obtained have been reviewed in Refs [10, 21].

4.1. Analysis by portable instruments

Shortly after X-ray probes and balanced filters had become commercially available, we made a field investigation of metallic-ore deposits in south

FIG.7. In-situ assay of Nb in pyrochlore-bearing vein at the Kvanefjeld plateau, south Greenland, by means of a portable X-ray fluorescence analyser. The probe contains a 1-mCi ^{109}Cd source and a Sr/Y balanced filter set.

Greenland [7]. In-situ assays of Nb in pyrochlore-bearing veins (Fig. 7), and of Zr, Nb, and La + Ce in rocks rich in eudialyte and rinkite were attempted using a 1-mCi ^{109}Cd source and a 3-mCi ^{241}Am source. A set of cut rock specimens with known contents of the elements to be assayed was used as calibration standards during the field work. There were no difficulties in finding clean and relatively smooth rock surfaces against which the X-ray probe could be held. Several hundred measurements were made in the course of two months. Metal contents assayed in the field were in close accordance with the contents found in chip samples taken at a number of measurement locations.

A similar investigation [10] was undertaken at the large, vein-type molybdenum deposit at Malmbjerget in east Greenland. With a spacing of 50 cm between the measurement locations, ore zones of varying grade were readily distinguished on the basis of in-situ X-ray fluorescence analysis. About 400 molybdenum assays were made in three days. Our general opinion is that portable X-ray fluorescence instruments are very suitable for direct assessment of metal contents in rocks, provided that these contents amount to at least $\sim 0.1\%$.

Field analysis of stream sediments has been tried during a traverse of a large drainage area some 20 km^2 in size in east Greenland [10]. The sediments were dried in the air and sieved, after which the -0.5 mm fractions were poured into a sample container and measured by means of an X-ray fluorescence analyser. We did not detect any Cr, Ni, Cu, Zn, Pb or Mo in the samples, whereas Zr and Nb were usually present in detectable amounts because zirconium-rich placer deposits are found within the drainage area. More elements might have been detected if heavy-mineral fractions had been separated and measured instead of measuring the sieved samples, but such a procedure would probably have resulted in serious matrix effects. Therefore, we do not recommend the use of portable X-ray fluorescence instruments for the analysis of stream sediments.

Following the same approach that has been used by several other investigators, we have studied X-ray fluorescence analysis of powdered, metallic ores [19]. A number of calibration standards have been prepared from a succesive dilution of minus-100 mesh fractions of chemically analysed ores with materials resembling the natural host rocks for the ores. For example, Cu, Zn, Zr, Pb and rare-earth minerals were mixed with powdered granite, and Cr and Ni minerals were mixed with powdered gabbro. For each ore metal, a curve was derived showing the difference count-rate versus the metal content.

4.2. X-ray spectrometry

In our first application of X-ray spectrometry we endeavoured to determine the abundances of rare-earth elements in cut rock specimens and sawed drill-core sections [20]. The samples were irradiated at an angle of 30° with ^{241}Am gamma rays, and spectra of the fluorescent radiation were recorded by means of a 3.5-mm-thick Ge(Li) detector and a multichannel analyser. The relative abundances of La, Ce, Pr and Nd were evaluated from a least-squares method based on the energies and intensities of the four K X-rays characterizing each element. An absolute calibration of the system was obtained by analysing 5-mm-thick layers of material removed from selected samples.

A high-resolution, 30-mm^2 Si(Li) detector was used in 1970. In a study reported in Ref. [9], a number of analysed rock powders, including some of the standards available from the US Geological Survey, were irradiated by means of a 3-mCi ^{109}Cd source. The intensities of the K X-ray peaks produced by Fe, Rb, Sr, Y, Zr, Nb and Mo were determined using a system of simultaneous equations. From the calibration data thus derived it was concluded that this sequence of elements could be assayed with an accuracy and precision of about 5%.

The analysis of stream sediments by X-ray spectrometry has just begun. For this purpose the K X-rays of the elements from potassium to selenium are excited by 30 mCi of ^{238}Pu. A least-squares data-reduction method is used in which allowance is made for the L X-rays produced by Pb in the samples and Au in the detector. A comparison between a measured X-ray spectrum and its calculated counterpart is given in Fig. 8. In the conversion of X-ray intensities into elemental abundances the intensities are normalized to the intensity of the backscatter radiation from the samples; in this way the matrix effect is greatly reduced.

5. BERYLLIUM ASSAY

Our development of beryllometers, operating on the basis of the $^{9}Be(\gamma, n)^{8}Be$ reaction, was actuated by the discovery in 1964 of a vein-type beryllium mineralization near the Kvanefjeld uranium deposit. The most abundant beryllium mineral is chkalovite (12% BeO), which is white and therefore difficult to distinguish from the gangue minerals.

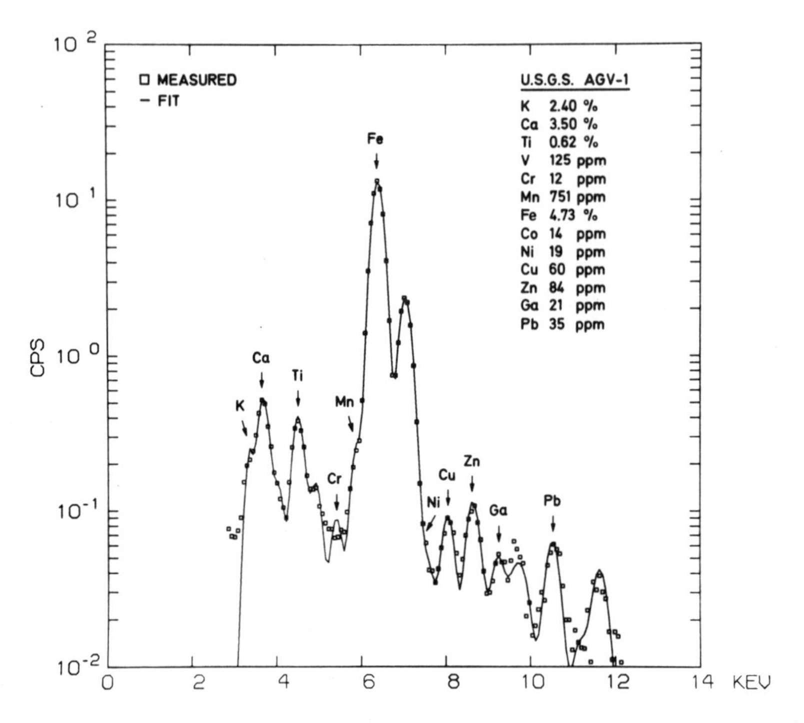

FIG.8. Measured Si(Li) X-ray spectrum of the US Geological Survey standard AGV-1, and the corresponding spectrum, determined from a least-squares fit.

The first field instrument we made [6] contained a single, 100-mCi ^{124}Sb source, mounted on a pivotal disk in a cylindrical lead housing. Two BF_3 counter tubes, embedded in paraffin wax, were used for detecting the photoneutrons released from the rock on which the instrument was placed. A robust calibration standard 20 cm × 20 cm × 5 cm and containing 2% BeO was prepared by dissolving $Be_4O(CH_3COO)_6$ in styrene and under-saturated polyester, which cured in an iron frame. Depending on the age of the source, the instrument produced a count-rate of 500 - 1000 counts/min per per cent of BeO in the rock, while the background count-rate was 1 - 3 counts/min. For a single field measurement of 10 minutes' duration the practical limit of detection was estimated to be 10 ppm BeO.

The veins on which the field assays were made had many surface irregularities and fissures produced by weathering. Since the count-rate of the beryllometer decreased by 15% per centimetre of average spacing between the vein material and the bottom of the instrument, some of the field readings were not considered trustworthy. Besides, the instrument was difficult to handle owing to its excessive weight (40 kg). These disadvantages were considered in the design of the second beryllometer constructed by us [5] (see Fig. 9). To reduce the distance effect, a 30-cm long array of ^{124}Sb sources was used instead of a single source. The neutron-detection efficiency was increased by the use of ^3He-filled counter tubes, pressurized to 4 atm, by which the total source strength, and thus the weight

FIG. 9. Field assay of beryllium on chkalovite-bearing veins in south Greenland. The beryllometer contains 20 mCi of ^{124}Sb, weighs 20 kg and has a sensitive area of about 500 cm^2.

of the lead housing, could be reduced. The improved performance in assaying densely mineralized veins was obtained at the cost of a higher detection limit.

Though none of the people who have carried and handled our beryllometers have received any impermissible radiation doses, we consider instruments like these to be rather hazardous.

6. FUTURE PROSPECTS

The Geological Survey has planned to cover as large a part of Greenland as possible by aerial, gamma-spectrometric measurements. Besides, radiometric follow-up work and geochemical sampling will be made in east Greenland and in other places where radioactive anomalies may be discovered. These activities suggest further studies regarding the interpretation of radiometric flight data recorded in mountainous terrains. The samples collected in the field will be analysed for uranium by delayed-neutron counting, and for ore metals by X-ray fluorescence spectrometry. By the end of 1975 a new, automatic counting set-up will go into operation at the 10-MW reactor DR 3 in which the samples will be irradiated in an almost pure thermal flux of 2.5×10^{13} n/cm$^2 \cdot$ s. Moreover, the Si(Li) X-ray spectrometer will be equipped with an automatic sample changer.

REFERENCES

[1] SØRENSEN, H., ROSE-HANSEN, J., LETH NIELSEN, B., LØVBORG, L., SØRENSEN, E., LUNDGAARD, T., The uranium deposit at Kvanefjeld, the Ilímaussaq intrusion, south Greenland, Geol. Survey of Greenland Report No.60 (1974).

[2] LØVBORG, L., KUNZENDORF, H., HANSEN, J., "Use of field gamma-spectrometry in the exploration of uranium and thorium deposits in south Greenland", Nuclear Techniques and Mineral Resources (Proc. Symp. Buenos Aires, 1968), IAEA, Vienna (1969) 197.

[3] LØVBORG, L., WOLLENBERG, H., SØRENSEN, P., HANSEN, J., Field determination of uranium and thorium by gamma-ray spectrometry, exemplified by measurements in the Ilímaussaq alkaline intrusion, south Greenland, Econ. Geol. 66 (1971) 368.

[4] LØVBORG, L., WOLLENBERG, H., ROSE-HANSEN, J., LETH NIELSEN, B., Drill-core scanning for radioelements by gamma-ray spectrometry, Geophysics 37 (1972) 675.

[5] LØVBORG, L., KUNZENDORF, H., HANSEN, J., "Portable beryllium prospecting instrument with large sensitive area", Nuclear Techniques and Mineral Resources (Proc. Symp. Buenos Aires, 1968), IAEA, Vienna (1969) 55.

[6] ENGELL, J., HANSEN, J., JENSEN, M., KUNZENDORF, H., LØVBORG, L., Beryllium mineralization in the Ilímaussaq intrusion, south Greenland, with description of a field beryllometer and chemical methods, Geol. Survey of Greenland Report No.33 (1971).

[7] WOLLENBERG, H., KUNZENDORF, H., ROSE-HANSEN, J., Isotope-excited X-ray fluorescence analysis for Nb, Zr, and La + Ce on outcrops in the Ilímaussaq intrusion, south Greenland, Econ. Geol. 66 (1971) 1048

[8] BOHSE, H., BROOKS, C.K., KUNZENDORF, H., Field observations on the kakortokites of the Ilímaussaq intrusion, south Greenland, including mapping and analyses by portable X-ray fluorescence equipment for zirconium and niobium, Geol. Survey of Greenland Report No.38 (1971).

[9] KUNZENDORF, H., An instrumental procedure to determine Fe, Rb, Sr, Y, Zr, Nb and Mo in rock powders by Si(Li) X-ray spectrometry, J. Radioanal. Chem. 9 (1971) 311.

[10] KUNZENDORF, H., "Non-destructive determination of metals in rocks by radioisotope X-ray fluorescence instrumentation", Proc. Fourth Int. Geochem. Explor. Symp. 1972, Inst. Min. Metall., London (1973) 401.

[11] LØVBORG, L., KIRKEGAARD, P., Response of 3" × 3" NaI(Tl) detectors to terrestrial gamma radiation, Nucl. Instrum. Methods 121 (1974) 239.

[12] LØVBORG, L., KIRKEGAARD, P., Numerical evaluation of the natural gamma radiation field at aerial survey heights, Risø Report No.317 (1975).

[13] LØVBORG, L., "Assessment of uranium by gamma-ray spectrometry", Uranium Prospecting Handbook, Inst. Min. Metall., London (1972) 157.

[14] LØVBORG, L., "Future development in the use of gamma-ray spectrometry for uranium prospecting on the ground", Uranium Exploration Methods (Panel Proc. Vienna, 1972), IAEA, Vienna (1973) 141.

[15] LØVBORG, L., LETH NIELSEN, B., "Processing and interpretation of radiometric flight data from central east Greenland", Geol. Survey of Greenland Report No. 55 (1973) 48.

[16] MOSE CHRISTIANSEN, E., SKAARUP, P., An airborne γ-scintillometer, Camac Bull. No. 6 (1973) 6.

[17] NATRELLA, M.G., Experimental Statistics, NBS Handbook No. 91, Washington (1963).

[18] CURRIE, L.A., Limits for qualitative detection and quantitative determination, application to radiochemistry, Anal. Chem. 40 (1968) 586.

[19] KUNZENDORF, H., LØVBORG, L., WOLLENBERG, H., Assay of powdered metallic ores by means of a portable X-ray fluorescence analyser, Risø Report No.251 (1971).

[20] KUNZENDORF, H., WOLLENBERG, H.A., Determination of rare earth elements in rocks by isotope-excited X-ray fluorescence spectrometry, Nucl. Instrum. Methods 87 (1970) 197.

[21] LØVBORG, L., KUNZENDORF, H., "Analysis of metallic ores by radioisotope-excited X-ray fluorescence", Risø Report No. 256 (1972) 85.

NUCLEAR TECHNIQUES OF ANALYSIS IN THE MINING AND MILLING OF BASIC METAL ORES IN AUSTRALIA

J.S. Watt
Australian Atomic Energy Commission,
Research Establishment,
Sutherland, New South Wales,
Australia

Abstract

NUCLEAR TECHNIQUES OF ANALYSIS IN THE MINING AND MILLING OF BASIC METAL ORES IN AUSTRALIA.

Analysis of basic metal ores is required at various stages of mineral exploitation. Nuclear techniques of analysis have little potential for use in the field at the early exploration stages, but some potential for down-hole and surface analysis when a mineral deposit is being evaluated before mining. The main potential for their use is in grade control at the mine face and for blending, and in continuous on-stream analysis of mineral process streams. Examples of nuclear techniques being developed or in use in Australia for grade control at the mine face or for blending are gamma-ray resonance scatter for copper and nickel ores, radioisotope X-ray techniques for tin, tungsten and iron ores, and neutron and selective γ-γ techniques for determination of iron in ore on a conveyor belt. Radioisotope on-stream analysis systems are in routine use in mineral concentrators for analysis of ore slurry streams.

INTRODUCTION

Quantitative analysis for basic metal elements in ores is required at various stages of mineral exploitation including exploration, evaluation of mineral deposits, mine operations and mineral processing. In assessing nuclear techniques for these applications, the real analytical requirement must first be defined. Then when considering available nuclear techniques, relevant factors include the accuracy required and the concentration ranges of the basic metal element to be determined, and the matrix elements, and the question whether measurements should be made in-situ and on-line or on representative samples. This choice in turn involves factors such as the in-situ conditions, particle size and time allowable for the analysis.

This paper outlines typical requirements for field and on-line analysis for the basic metal content of ores at the various stages of mineral exploitation, and describes some of the nuclear techniques of analysis being developed and in routine use in Australia.

EXPLORATION AND EVALUATION OF MINERAL DEPOSITS

The principal mineral occurrences in Australia are closely related genetically to their stratigraphic environment, and mineralisation often extends or recurs over distances measured in kilometres and even tens of kilometres [1]. Concentrations of basic metals in the mineral field often vary from economic grade to very low levels which are still above average for the region and can indicate possible mineralisation. Most of the important mineral deposits discovered in Australia over the last twelve years have had weak or unrepresentative surface expression, or with outcrops of grade which would not have attracted the earlier prospector [2].

At the exploration stage aimed at locating a mineral field, nuclear techniques of analysis have little potential except for the few very simple ones such as natural γ-ray measurements for uranium, thorium and potassium. Once mineralisation has been located, surface samples and (at a later stage) drill cores are analysed for as many metallic and related elements as possible and at concentrations down to levels of a few parts per million. These analyses are best undertaken in a central laboratory using atomic absorption spectrometry and X-ray fluorescence (tube plus Bragg crystal spectrometer) techniques, and, in a limited number of cases of precious metals, using neutron activation analysis. This laboratory assay is outside the scope of this paper.

Once a significant mineral deposit has been located it must be evaluated in detail which includes location of the boundaries of the deposit, assessment of the amount of ore and average grade, and determination of structural information for mine planning and mineralogy for design of mineral concentrators. Most of this information is obtained from tests on sample cores from drill holes. Since cores are necessary for much of this evaluation there is little point in introducing down-hole nuclear techniques of analysis because this is usually more simply accomplished on the cores in the laboratory. The only justification for use of nuclear techniques of analysis down-hole would be when this is less expensive than laboratory assay, or satisfactory core samples could not be obtained, or it can be demonstrated that the larger amount of ore 'seen' by the nuclear technique gives necessary information unobtainable from assay of the core.

However, at the later stages of evaluation, information on mineralogy and structure is not required from every drill hole. Down-hole logging for concentration of basic metals could be made in percussion holes which are usually less expensive to drill. Sensitivity of analysis required is often about 10% of the cut-off grade of economically minable ore and for many basic metals this sensitivity is several hundred parts per million or higher. Generally the sensitivity of nuclear techniques is sufficient (or approaches it) for this down-hole analysis at the evaluation stage, but further development is still required to overcome effects of borehole parameters and to provide better calibration for element concentration. Most of the development in Australia has been related to analysis for iron [3], nickel [4,5] and copper [4] but the methods have not yet been put into routine use.

There has been far less interest, in Australia at least, in nuclear techniques for determining metallic element content of ores at the exploration and evaluation stages than for the stages of mine operations and control.

MINE OPERATIONS

Analysis of ore is required for grade control at the mine face and for blending prior to processing. The main need is to determine the concentration of valuable metallic element(s) averaged over very large amounts of ore. Economic grades for basic metals range from several tenths of a wt.% for elements such as copper, nickel, tin and tungsten, several wt.% for lead and zinc, and several tens of wt.% for iron and aluminium.

The most important decision to make in grade control is whether the mined rock is above or below economic cut-off grade; that is whether to process it as ore or dump as waste. With improved grade control, the recovery of ore is increased and/or the uneconomic processing of waste reduced. Determination of the concentration of the basic metal element to within 20% of the mean throughout the given ore block is usually quite sufficient accuracy for grade control, and in practice most of this error will be due to difficulties in obtaining a representative sample. In many cases the analysis to determine whether the ore is above or below economic cut-off grade is most simply achieved after the blast holes have been drilled. In open cut mines the requirement for analysis is usually ore grade averaged over the whole depth of the blast hole, and not variations with depth. The simplest method of determining this average grade is sampling from the drill chippings and analysis by nuclear techniques. In underground operations, information on ore grade with depth in the blast hole is more often required as in some cases blasting can be limited to the boundary of economic cut-off grade ore. In-situ logging by nuclear techniques is hence of more importance in underground mining.

Blending is usually undertaken to obtain a more uniform grade of ore and, in some cases, impurities. The most important example in Australia is the blending of iron ores in the Pilbara region of West Australia.

Copper and nickel ores

Grade control for copper and nickel ores is most simply achieved by γ-ray resonance scattering [6]. This technique depends on the selective scattering of the γ-ray when it is in exact resonance with a nuclear energy level of the element of interest; it can be made highly specific to either copper or nickel. The resonance scatter analyser (Fig.1) has been fully tested in the laboratory on a wide range of copper ore samples [7]. The resonance scatter count rate is proportional to copper concentration (Fig.2), and accuracy of analysis is improved by making a small correction for density of the ore. Copper at a concentration of about 1 wt.% can be determined to ±0.05 wt.%[1] in about 100 seconds. This analytical sensitivity is sufficient for almost all applications at the mine face. The analysis is averaged over about 50 kg of sample.

Analysis of drill chippings in the field with the resonance scatter analyser could be used for grade control at open cut mine operations such as Bougainville Copper Pty. Ltd. and Kanmantoo Mines Ltd. where the grade of ore mined is in the range 0.3 to 1 wt.% copper. This would overcome the present need for sample splitting, grinding and laboratory assay, and provide faster analysis on larger samples, thus ensuring savings in time and manpower as well as analysis of a more representative sample.

At Mount Isa Mines Ltd. a somewhat different problem is encountered. The mining is underground and the stopes are separated by pillars of ore. After these stopes are worked out they are filled with a mixture of siltstone and cemented hydraulic fill. The pillars are then mined. After blasting it is very difficult to assess ore grade near the interface of ore and fill because these become mixed. A resonance scatter analyser mounted in a vehicle near the drawpoint would provide a rapid analysis averaged over large sample weights and hence greatly improve grade control. The resonance scatter analyser is being field tested now (November 1974) at Mt. Isa Mines on samples taken from the drawpoints [7].

Ores of high atomic number elements

Grade control for high atomic number elements is most simply accomplished using radioisotope X-ray techniques of analysis depending on K shell fluorescence, and in some cases, by backscatter or preferential absorption of low energy γ-rays. Basic metal elements of interest are tin, barium, tungsten and lead.

[1] All errors quoted in this paper are one standard deviation.

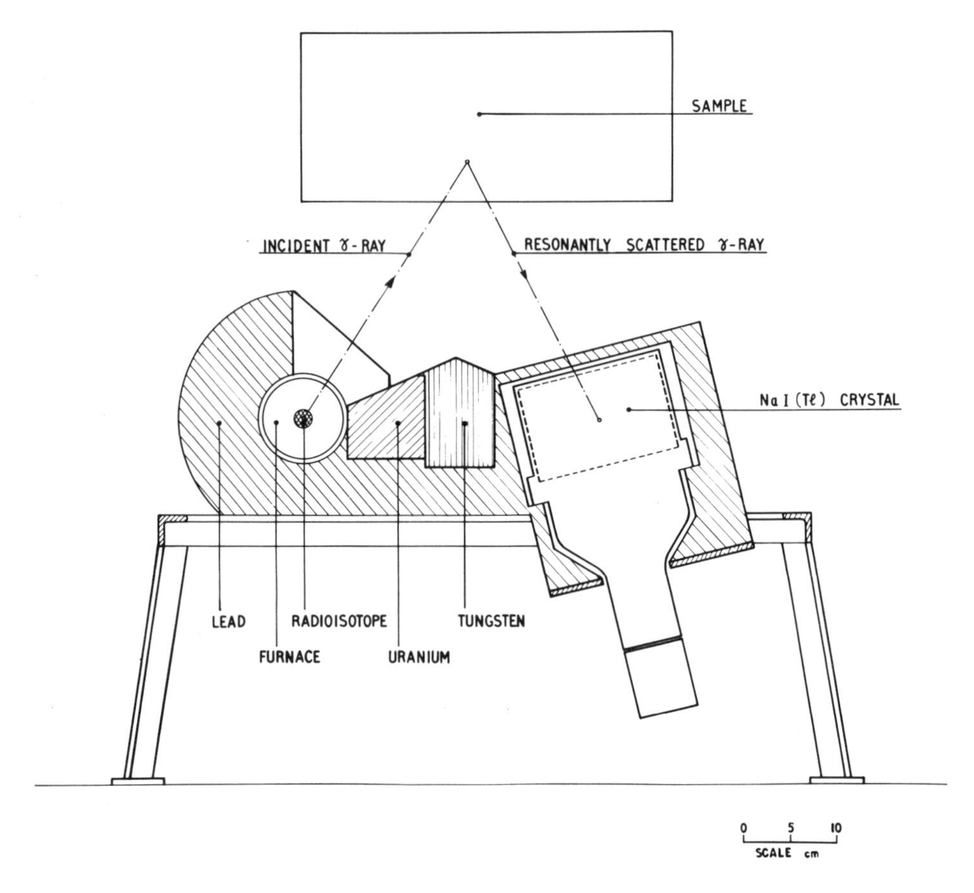

FIG. 1. Gamma-ray resonance scatter analyser for determination of copper and nickel in ores.

At Renison in Tasmania a finely disseminated tin ore is mined. Long-term control of the underground mine is based on distribution and grade of tin predicted using assay data from drill cores and tight control of grade between the diamond drill holes is achieved by using data on assays for tin in blast holes. This is obtained by in-situ logging using radioisotope X-ray techniques based on balanced filters [8]. The in-situ assays define the boundary of ore at economic cut-off grade, and if a large block of ore is below economic limits it is blasted and left in the stope.

Radioisotope X-ray techniques using a portable mineral analyser were tested at the Mt. Cleveland mine, Tasmania, for determination of tin at the mine face [9]. Although results compared well with conventional sampling techniques, and there appeared to be considerable savings in cost and time, the method is no longer used [10].

Grade control for tungsten in scheelite ores is required at the underground mines of King Island Scheelite. Blast holes are drilled into the mine face using percussion techniques, and it is very difficult to

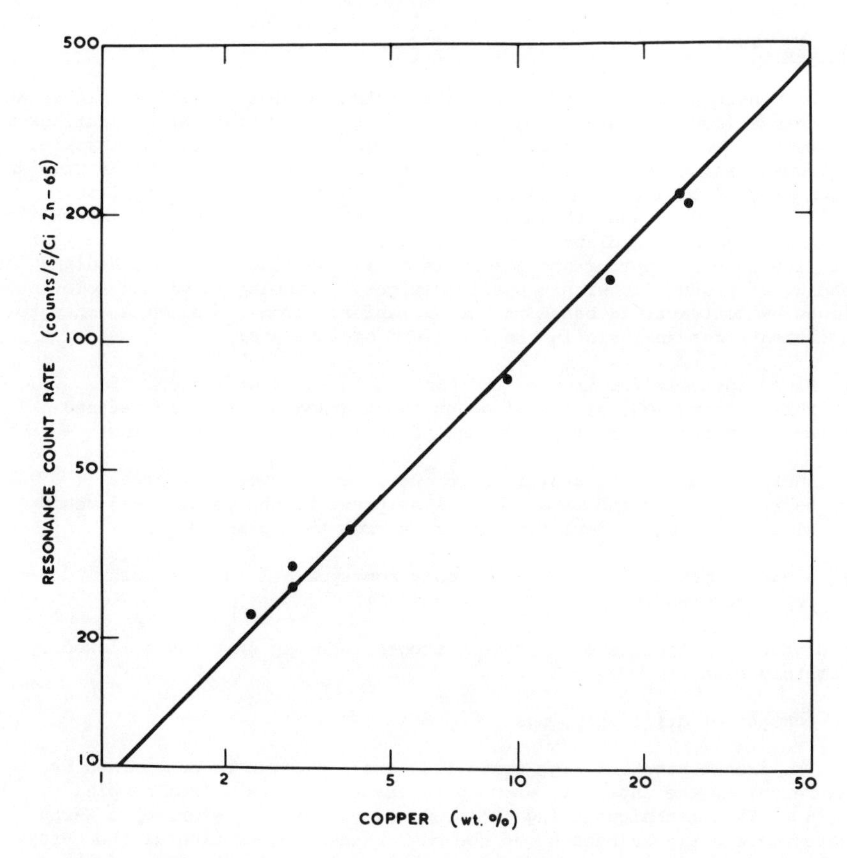

FIG. 2. Resonance scatter count-rates for samples of ores, concentrates and mixtures of powdered materials having densities in the range 1.3 to 3.2 g/cm^3. The count-rate is zero at 0 wt.% copper. Accuracy of determination of copper is improved by correcting for density of the sample.

collect a sample representative of the ore from the holes because the scheelite ore particles are very fine and mixed with much coarser fragments of rock and large amounts of water. In-situ analysis in the blast holes appears the best approach to determination of ore grade, and analytical sensitivity required is about ±0.03 wt.% tungsten at the economic cut-off grade of 0.25 wt.%. Radioisotope X-ray techniques are being developed for this in-situ analysis, and depend on K shell fluorescence of tungsten excited by cobalt-57 γ-rays [11].

Although K shell fluorescence and γ-ray backscatter techniques have been developed to determine lead in samples and in-situ in boreholes [12,13], these techniques have not been used in Australia. Economic cut-off grade not only for lead but also for zinc is required and the only simple nuclear technique for zinc (K shell fluorescence) is not satisfactory because of the limited sample weight of ore 'seen' by the X-rays. There is a real need for a better nuclear technique of analysis for zinc.

Iron Ore

As mentioned above, in the Pilbara region of West Australia, higher and lower grade iron ores are blended to give a product within specifications of iron content and with an acceptably low level of impurities. For example, Mt. Newman Mining Co. mines ore which has a mean iron content of 64 wt.% and ranges from 54 to 70. The operation [14] is based on having negligible storage capacity between the mining face at Mt. Whaleback and the stockpile 400 km away at Port Hedland. Grade control at the mine face ensures that sufficient ore of appropriate grades is always available at Port Hedland for blending of product to within specifications. Blending to within tolerances allowed by contracts is based on two variables: iron and alumina contents. Requirements for analysis in the field and on-line are:

(a) Field analysis for iron at the blast hole to ±1 wt.% iron. From this information, blocks of about equal grade of iron are defined and form the basis of scheduling of excavation after blasting.

(b) Analysis for iron and alumina in ore on a conveyor belt after coarse crushing and prior to rail shipment to the port. This would give a warning of deviation in grade from that planned.

(c) Accurate determination of iron on a conveyor belt at the port prior to blending.

All of these analyses are at present accomplished by sampling followed by laboratory assay.

Iron content of drill chippings

The iron content of high grade iron ore can often be determined by measurement of the intensity of Compton scattered γ-rays from samples [15,16]. This technique using ^{241}Am 60 keV γ-rays is preferred to X-ray fluorescence analysis because of the much higher penetration of the γ-rays and hence analysis is averaged over much larger sample weights and the effect of particle size is much reduced. The immediate application of interest is field measurement of the iron content of drill chippings from blast holes at Mt. Newman Mining Co., and this should be achieved simply by using an instrument of the portable mineral analyser type.

Laboratory measurements made on ore samples supplied by Mt. Newman Mining Co. showed that iron in drill chippings could be determined to about ±1 wt.% for iron in the range 40 to 68 wt.% and iron in finely ground ore to ±0.13 wt.% over the range 57 to 67 wt.% [16]. These results indicate the potential for field measurement of iron in blast hole drill chippings but further field tests will have to be undertaken to fully prove the technique.

On-line analysis of iron ores

The Commonwealth Scientific and Industrial Research Organisation has developed two methods for determination of the grade of hematite iron ores under bulk handling conditions.

The first method [17] has been developed for ore on natural rubber conveyor belts where crushed ore is exposed to an isotopic neutron source under the belt and the resultant thermal neutron flux and iron capture gamma ray flux are monitored (Fig.3). Depth of ore on the belt is monitored independently, and grade is calculated from the observed signals by means

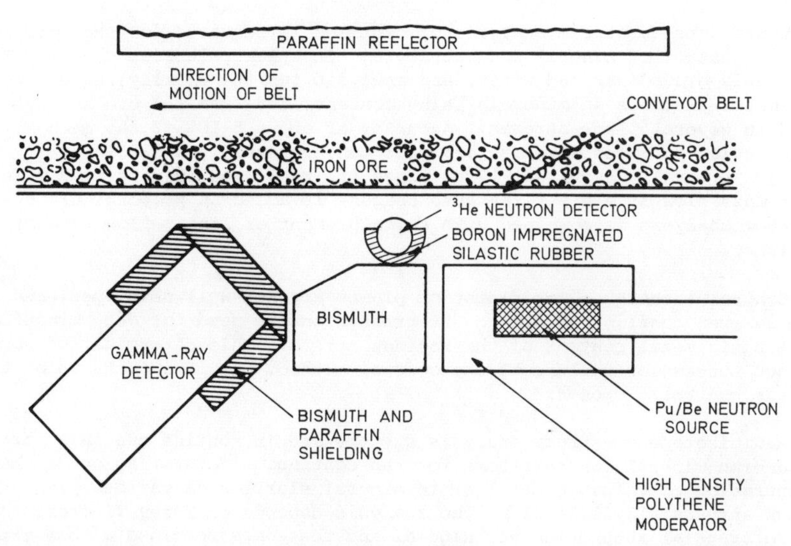

FIG. 3. Analysis system for determination of iron in iron ore on a conveyor belt using neutron capture gamma-ray techniques.

of a regression equation. Calibration with known grades of material is required to determine the regression coefficients. Accuracy is better than ±1 wt.% iron at the 63% iron level.

The second method under development is an "above the sample" technique involving measurement of back-scattered ^{60}Co radiation and the Blumencev-Czubek "P-" factor, first developed for borehole logging [18,5]. This factor varies almost linearly with grade and permits batch analysis of bulk iron ore samples at an accuracy of ±0.25 wt.% iron at the 63% iron level. Second-order corrections are made for changes in bulk density. Crushing is not needed unless highest accuracy is essential. Conveyor belt application requires either that the depth of ore be maintained constant, or that the depth be monitored and included in the regression equation for grade determination.

MINERAL PROCESSING PLANTS

The mineral industry has had a need for continuous analysis of ores on conveyor belts, and continuous on-stream analysis of process streams in mineral processing plants for many years. Analysis of ores on conveyor belts can make possible the separation of ore into several grades (e.g. high, medium and low), with each grade routed to a separate storage bin. Continuous feed from one bin would then provide relatively constant grade input to the processing plant over relatively long periods allowing better optimisation and less frequent changes of plant control parameters. In the past, analysis of process streams involved sampling followed by laboratory assay which was slow, and results would often be available too late for control purposes. With the development of continuous on-stream analysers there is much less need for sorting prior to processing, and ore sorting is apparently no longer a significant need.

Continuous on-stream analysis of slurries is required for the more economic control of mineral concentrators. The plant slurries consist of very finely ground ore and water, and analysis for the usually one or two basic metal elements in minerals being concentrated from the ore is required in several plant streams. Accuracy of about 5-10% of the mean concentration for feed streams, 1-5% for concentrates, and 5-20% for tailings is usually required [19]. Since the real need in most feed streams is for mass flow of element (because reagent is added in proportion), the on-stream analysis is combined with a measurement of linear flow rate of the slurry.

Control techniques for flotation plants are not well developed, and there is even conflict of opinion at present on the need for determination of the basic metal content of the residue stream. This determination would indicate success or failure of the control method, but it appears to be too late for control purposes.

Radioisotope on-stream analysis systems are in routine use in Australian and Zambian mineral concentrators for the continuous determination of the concentration of valuable metal(s) in mineral slurries of various plant process streams [19,20,21,22]. The analysis depends on X-ray fluorescence and preferential absorption techniques, and these are combined with a γ-ray absorption measurement for density of the slurry to give concentration of element either per unit weight of slurry or of slurry solids [23].

Analysis for elements of atomic number about 25 (manganese) and above can be made with the radioisotope on-stream analysis system. Feasibility studies have been successfully undertaken on many different types of ore, analysing for iron, nickel, copper, zinc, zirconium, tin, lead, bismuth and uranium. The minimum detectable level depends on atomic number of element and type of ore, but for slurries it is usually in the range 0.01 to 0.03 wt.% in the slurry solids.

Probes containing radioisotope X-ray source, scintillation detector and preamplifier are immersed directly into plant slurry streams (Fig.4) and signals are fed to a prescaler unit located close to the probes in the

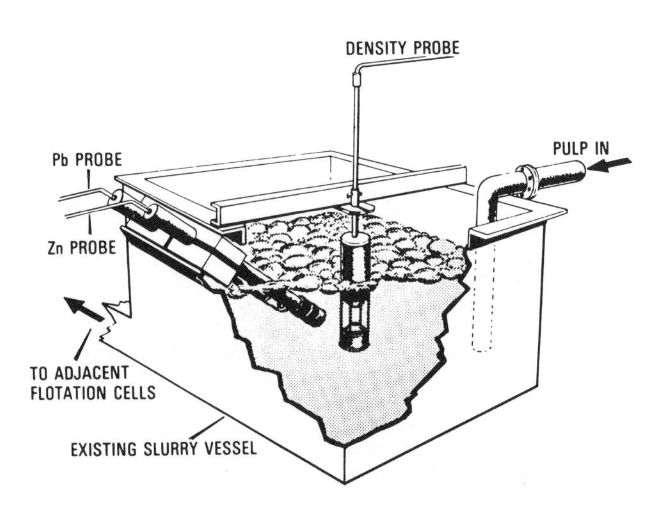

FIG. 4. Arrangement of radioisotope immersion probes in plant stream.

TABLE I. PLANT INSTALLATIONS OF RADIOISOTOPE ON-STREAM ANALYSIS SYSTEMS

Plant	Analysis for	Number of streams analysed
North Broken Hill Ltd.	lead	1
New Broken Hill Consolidated Ltd.	lead	2
	zinc	3
Kanmantoo Mines Ltd.	copper	3
Cobar Mines Ltd.	copper	3
Mt. Lyell Mining and Railway Co.	copper	2
Roan Consolidated Mines Ltd. (Chambishi Mine, Zambia).	copper	3

plant [21]. Signals from the prescaler unit are fed to a small computer which calculates the instantaneous concentration of valuable metal in the stream. The advantages of this system are that there is no need for sampling of slurry streams, each stream is continuously analysed, the overall system is relatively inexpensive, and it is possible to build up the plant system stream by stream as the need for on-stream analysis develops.

Plant installations of the radioisotope on-stream analysis system are listed in Table I. The average number of streams analysed per plant is about three, but this will no doubt increase because in some plants selective flotation for more than one valuable element is required. Proof of user confidence in the system is indicated by the recent doubling of the probe installation at New Broken Hill Consolidated Ltd. after three years' experience with the initial installation, and plans by another user company to increase the capacity of their on-stream analysis system.

Experience has shown that calibration of the probes is maintained over long periods. For example, at Mt. Lyell Mining and Railway Co. probes to determine copper in the flotation feed stream were calibrated, and over the next four month period copper concentration determined by the on-stream system was compared with conventional assay of samples taken from the stream (Fig.5) [24]. The correlation between on-stream results and assay of samples is obviously good. The accuracy for copper, determined by comparison with assay of samples, was ±0.05% copper by wt. in the slurry solids. This includes errors due to sampling and conventional assay.

Another example of checks of on-stream analysis by comparison with assay of samples taken from the stream is shown in Fig.6 [25]. These results from the New Broken Hill Consolidated plant show excellent agreement between the two methods for zinc assay. The company reports that the probes have been operated for three years without significant mechanical or electrical fault [26].

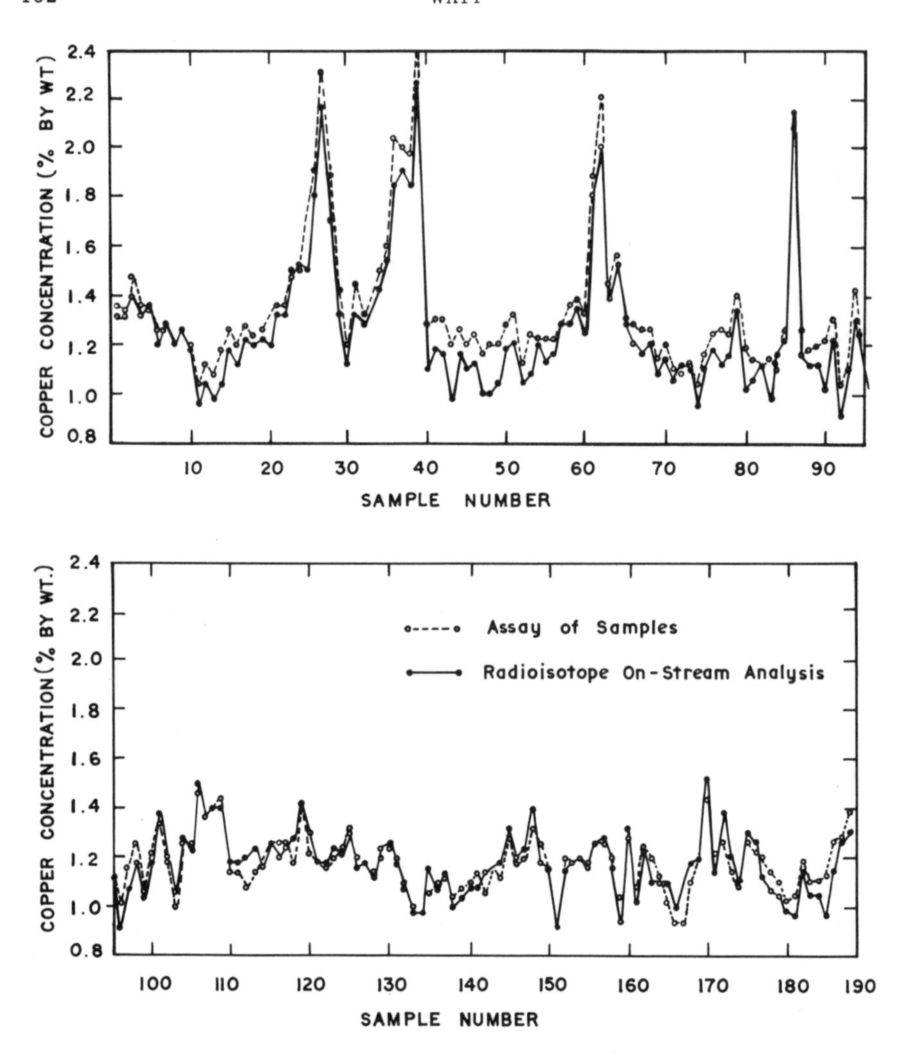

FIG. 5. Determination of copper in the flotation feed stream at the Mt. Lyell plant showing four months of results of radioisotope on-stream analysis and assay of samples taken from the stream.

Solid-state detectors can be used for on-stream analysis if the need for determination at very low concentrations arises. For example, copper in residue streams can be determined to ±0.004 wt.% [27]. The best approach at present is to use a short sample byline from the main stream to the detector mounted in the plant.

The main need for continuous on-stream analysis is for improvement of control of mineral concentrators. Research and development into better control techniques based on signals from radioisotope probes and other sensors is now being undertaken by various companies and organisations in Australia. A simple control loop has been incorporated at the plant of

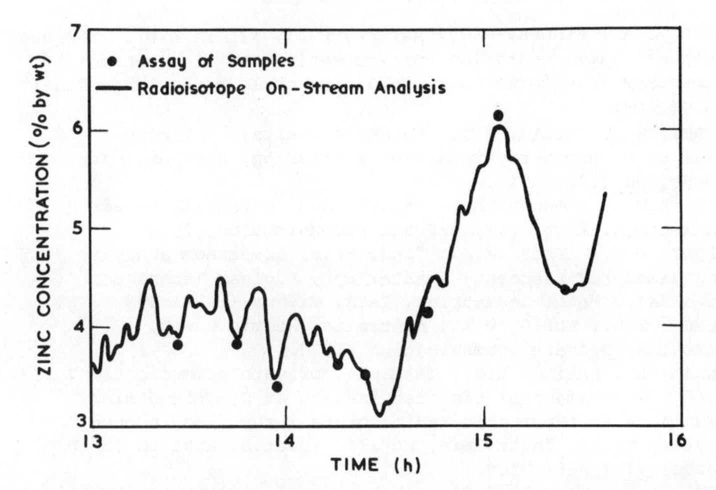

FIG. 6. Radioisotope on-stream analysis record for zinc in the zinc rougher tailing stream of the New Broken Hill Consolidated plant showing check assays of samples taken from the stream.

New Broken Hill Consolidated Ltd. which ensures that flotation reagent is added in proportion to mass flow of lead into the flotation circuit [25]. The mass flow of lead is determined by combination of signals from the radioisotope probes and a magnetic flow meter.

A new approach to elucidation of the dynamics of flotation circuits is being investigated by the Australian Atomic Energy Commission and is based on fluctuation analysis of signals from radioisotope probes and other sensors [28]. Cross-correlations between certain input and output signals of the lead flotation circuit have been established using tapes of signals from the New Broken Hill Consolidated plant. Considerably more work is necessary before the value of fluctuation analysis in control of mineral concentrations is established.

ACKNOWLEDGEMENTS

The author thanks the following for supply of information used in this paper: Dr. B. D. Sowerby and Mr. W. K. Ellis, Australian Atomic Energy Commission,for results of recent tests of the resonance scatter analyser including Figs. 1 and 2.; Dr. A. W. Wylie, Commonwealth Scientific and Industrial Research Organisation,for the description of the on-line determination of iron in iron ore including Fig.3; the Australian Mineral Development Laboratories, New Broken Hill Consolidated Ltd., and Cobar Mines Pty. Ltd. for supply of details relating to installations of radioisotope on-stream analysers. The assessment of the field and plant analytical requirements in the mineral industry has resulted from continuing discussions by Dr. Sowerby and myself with people in the Australian mineral industry.

REFERENCES

[1] KING, H.F., Some antipodean thoughts about ore, Economic Geology, 68 (1973) 1369.

[2] KING, H.F., "A look at mineral exploration 1934-1973", 1973 Annual Conference-Technical Papers, Australas.Inst.Min.Metall., Melbourne (1973) 1.

[3] AYLMER, J.A., EISLER, P.J., MATHEW, P.J., WYLIE, A.W., The use of natural gamma radiation for estimating the iron content of sedimentary iron formations containing shale bands, This Panel Proceedings.

[4] SOWERBY, B.D., ELLIS, W.K., Borehole analysis for copper and nickel using gamma-ray resonance scattering, Nuclear Instr. Methods, 94 (1973) 511.

[5] WYLIE, A.W., Commonwealth Scientific and Industrial Research Organisation, Australia, private communication, 1974.

[6] SOWERBY, B.D., ELLIS, W.K., "Industrial on-stream analysis using gamma-ray resonance scattering", Nuclear Techniques in the Basic Metal Industries, IAEA, Vienna (1973) 479.

[7] SOWERBY, B.D., ELLIS, W.K., Australian Atomic Energy Commission, private communication, 1974.

[8] NEWNHAM, L., Renison Ltd., Tasmania, private communication, 1974.

[9] COX, R., Assessment of tin ores in-situ at Cleveland mine, Tasmania, with a portable radioisotope X-ray fluorescence analyser, Trans. Instr. Min. Metall. (Sect.B: Applied Earth Science) 77 (1968) B109.

[10] MURPHY, A.J., Cleveland Tin N.L., Australia, private communication, 1974.

[11] GRAVITIS, V.L., GREIG, R.A., WATT, J.S., Australian Atomic Energy Commission, private communication, 1974.

[12] MARR, H.E., CAMPBELL, W.J., Evaluation of a radioisotopic X-ray drill-hole probe, U.S. Dept. of the Interior, Bureau of Mines report RI7611 (1972).

[13] CHARBUCINSKI, J., A selective non-spectrometric gamma-gamma method of determining the zinc and lead content of polymetallic ores, Nukleonika XV 7-8 (1970) 563.

[14] SMITH, G.G., "Quality control at Mt Newman", 1973 Annual Conference - Technical Papers, Australas. Inst. Min. Metall., Melbourne (1973) 323.

[15] DZIUNIKOWSKI, B., NIEWODNICZANSKI, J., "Field determination of iron using X-ray fluorescence and γ-ray scattering", Nuclear Techniques and Mineral Resources, IAEA, Vienna (1969) 343.

[16] FOOKES, R.A., GRAVITIS, V.L., WATT, J.S., Determination of iron in high-grade iron ore and of lead in lead concentrate by Compton scattering of 60 keV γ-rays from americium-241, Anal.Chem. (in press).

[17] HOLMES, R.J., McCRACKEN, K.G., WYLIE, A.W., Analysis utilising neutron irradiation, Australian Patent Appl. PA9294 (1972).

[18] CZUBEK, J.A., "Physical possibilities of gamma-gamma logging", Radioisotope Instruments in Industry and Geophysics, Vol.II, IAEA, Vienna (1966) 249.

[19] WATT, J.S., HOWARTH, W.J., "Mineral processing methods and on-stream analysis in Australian mineral processing plants", Nuclear Techniques in the Basic Metal Industries, IAEA, Vienna (1973) 105.

[20] WATT, J.S., Radioisotope on-stream analysis, Atomic Energy in Australia 16 4 (1973) 1.

[21] PHILIPS SCIENTIFIC AND INDUSTRIAL EQUIPMENT, Radioisotope on-stream analysis, Sydney (1974).

[22] HOWARTH, W.J., WENK, G.J., WILKINSON, L.R., "Installation and operating experience with radioisotope on-stream analysis equipment", Review of On-Stream Analysis Practice, Australian Mineral Industries Research Association Ltd., Melbourne (1973) 107.

[23] WATT, J.S., GRAVITIS, V.L., "Radioisotope X-ray fluorescence techniques applied to on-stream analysis of mineral process streams", IFAC Symposium on Automatic Control in Mining Mineral and Metal Processing, Institution of Engineers, Australia, National Conference Publication No. 73/4 (1973) 199.

[24] WENK, G.J., HARTLEY, D.G., TOMPSITT, D., Note on performance of Mt Lyell on-stream analysis system, CIM (Can. Min. Metall.) Bulletin 67 748 (1974) 118.
[25] STUMP, N.W., ROBERTS, A.N., On-stream analysis and computer control at the New Broken Hill Consolidated Ltd. concentrator, SME preprint 74B34, AIME Annual Meeting, Dallas, Texas (1974).
[26] HARDWICK, W.S., New Broken Hill Consolidated Ltd., Australia, private communication, 1974.
[27] GRAVITIS, V.L., GREIG, R.A., WATT, J.S., X-ray fluorescence analysis of mineral samples using solid state detector and radioisotope X-ray source, Proc. Australas. Inst. Min. Metall. 239 (1974) 93.
[28] JOZSA, R.O., LAWRENCE, B.R., HARRIS, R.W., "Feasibility of fluctuation analysis for flotation dynamics identification", Optimisation and Control of Mineral Processing Plants, Australian Mineral Industries Research Association Ltd., Melbourne (1974) 91.

PORTABLE X-RAY AND ON-LINE NUCLEAR INSTRUMENTATION

Present status of applications in the mining industry in the United States of America

J. R. RHODES
Columbia Scientific Industries,
Austin, Texas,
United States of America

Abstract

PORTABLE X-RAY AND ON-LINE NUCLEAR INSTRUMENTATION: PRESENT STATUS OF APPLICATIONS IN THE MINING INDUSTRY IN THE UNITED STATES OF AMERICA.

The current status of X-ray and neutron instrumentation for field and process analysis in the USA mining industry is described, with special reference to applications in exploration, development, mine control and process control of mineral beneficiation.

1. INTRODUCTION

Developments in radioisotope X-ray and neutron techniques and instrumentation for field analysis and for process control have been well documented in recent I.A.E.A. publications [1-4]. The advantages and economic benefits of applying this technology in specific segments of the mining and metallurgical industries have also been discussed [1,3,4].

In spite of this information being available, applications in the USA Mining Industry have spread slowly, and until recently were largely restricted to a few "leading" companies. The purpose of this paper is to report on the present status of these applications in exploration, development, mine control and process control of hard minerals.

2. PORTABLE X-RAY ANALYZERS

2.1. Standard Portable XRF Analyzers

Portable XRF analyzers using a radioisotope source for excitation and balanced filters for energy selection have been commercially available in the USA since 1968. Of the four companies that sell this instrumentation two serve the Mining Industry. Sales to the Mining Industry represent about 70% of one company's [5] total portable analyzer sales and about 25% of the other's [6]. The total number of instruments sold for mineral analysis, estimated at 125, represents roughly one-third of the total portable analyzer sales by the four companies. Although initial evaluations of field applications began in 1967, most of the above-mentioned sales have occurred in the last three years and the rate of sales is still increasing.

Both instruments in use in Mining are based on a NaI(Tl) scintillation probe, single channel analyzer and digital readout. In one case a side-source geometry is used [6] and in the other, the standard central source geometry is retained but other variations such as a multifilter head [5] are offered.

Individual applications are hard to trace because many instruments after purchase by a US-based company are transported to different locations, often outside the USA, and used for a variety of analyses. Most of the applications are for single element determination in mine control. These comprise the following ore analyses: vanadium, iron, copper [7], molybdenum, tungsten, uranium (to 5 to 10 instruments each); chromium, manganese, zirconium, niobium, silver and tin (1 to 2 each); and ash in coal (X-ray backscatter, one instrument). The remainder are in use for two-element determinations (Ca and Fe in cement raw mix and in iron ore sintar mix; Ti and Fe in clays; and Ti and Zr in sands) and for multielement analyses. The major multielement applications are for determination of Cu, Zn and Pb in Cu-Zn-Pb ores [8] and Mn, Fe, Ni and Cu in manganese nodules [9]. All US-based deep-sea mining operations are using the portable analyzer for shipboard nodule analyses. There appears to be an increasing trend towards multielement applications.

2.2. Variations on the Standard Equipment

A 4-cm diameter borehole logging probe using two scintillation detectors, sources and filters was developed for the U. S. Bureau of Mines and was originally evaluated in the laboratory for silver and uranium determination [1]. Subsequent field tests were conducted by the Bureau of Mines for lead ore assay using ^{57}Co to excite Pb K X-rays and W and Au balanced filters to isolate them [10]. Although the trials indicated that the probe response correlated reasonably well with ore grade, the project was shelved.

A second special probe was developed for determination of cement content of fresh concrete [11]. The intensity of scattered gamma-rays from ^{241}Am is used as a measure of the cement-silica ratio. The results of simulated field evaluations [12] of the prototype instrument indicate that the method is suitable for certain concrete analyses, especially if the aggregate is siliceous, that is, contains no limestone. Two units are at present undergoing field evaluation. Neutron techniques have also been developed for concrete analysis (see Section 3.1.2). Comparative evaluation of all equipment for field quality control of fresh concrete is to take place in 1975.

A drill core analyzer has been developed and tested for assay of copper and zinc in whole cores [6]. The instrument uses a standard portable XRF analyzer coupled to a small lathe. Scanning speed and pitch, and signal integration time are independently adjustable. In an evaluation, one to three foot lengths of 1-11/16 in. diameter core were scanned in three-inch segments, taking 60 seconds per measurement. The segments were subsequently pulverized and assayed by atomic absorption and X-ray fluorescence (using the portable analyzer). Plots of both X-ray signals vs atomic absorption assay for corresponding samples showed excellent linear correlation for both copper and zinc [6]. Table I lists the standard deviations of the calibration curves obtained. As with the drill hole probe, this clear proof of practicability and economic benefit has not yet resulted in significant commercial application.

TABLE I. RESULTS OF CORE ANALYZER EVALUATION FOR Cu AND Zn IN WHOLE CORES

	% Cu	% Zn	
Powder Samples XRF vs AA	0.09	0.07	Std. Dev. % Element
Core Surface Scan XRF vs AA	0.23	0.12	Std. Dev. % Element
Concentration Range	0 to 2	0 to 3	

Total number of samples in each regression analysis, 57

2.3. Relevant New Technology

A number of compound semiconductors are being investigated as detectors for room temperature, high resolution X-ray spectrometry, and progress up to 1973 was reviewed by Langheinrich et al. [4]. Recent work on HgI_2 shows special promise. Resolutions of 850 eV (fwhm) at 5.9 keV and 4.3 keV at 122 keV have been obtained using detectors up to 1 mm in thickness and a few mm^2 in area [13]. Charge carrier transport characteristics and crystal growth conditions are under active investigation with good prospects of producing larger detectors having better resolution.

A portable XRF analyzer incorporating a cryogenically-cooled Si(Li) spectrometer is now offered commercially [14] but the results of field evaluations are not yet available.

A space-hardened XRF spectrometer, employing proportional counters and balanced filters, was developed and flown in lunar orbit in the Apollo 15 and 16 flights [15]. Mg to Mn K X-rays excited in the lunar surface layers by incident solar radiation were measured and Mg/Si and Al/Si concentration ratios mapped.

Another space-hardened instrument has been developed and tested for the 1975 Viking Mars Lander [16-18]. It comprises ^{109}Cd and ^{55}Fe sources, each coupled to a pair of proportional counters, the gas filling and window material of which are chosen judiciously to enhance spectral discrimination. The objective is rock-type analysis on Mars. Preliminary tests showed that the instrument could determine the following elements with satisfactory sensitivity and accuracy: Mg, Al, Si, K, Ca, Ti, Fe, Rb, Sr, Zr and "total light elements" (Z = 1 to 11). Raw data are transmitted to a ground station and reduced by an iterative procedure which starts with a rough composition estimate based on standard rock spectra.

3. PROCESS ANALYZERS

3.1. Neutron-Gamma Instrumentation

In the past four years a significant increase in the rate of feasibility studies has occurred through the USAEC Cf-252 Evaluation Program. The work reported here is limited to instrument developments relevant to Mining which have reached the stage of extended plant trials or commercial availability.

The most recent ^{252}Cf market survey [19] shows process control instrumentation to be the largest contributor to the potential market. For the period 1974 to 1980 it is estimated that some 250 such instruments will be sold for a value of about US $16 M. Sales to date are thought to be approximately US $1 M, in line with earlier predictions. It is likely that the participants in the market survey made no distinction as to neutron source type, so the above figures are also the best estimates for all process analyzers based on neutron-gamma techniques.

3.1.1. On-Stream Analyzers

Both government and private industry are active in the USA in development of process analyzers for bulk materials. The first commercial application was realized in 1970 with the commissioning of an on-stream taconite analyzer for silicon determination [20]. Since then it is estimated that three more similar instruments have been installed in iron ore beneficiation plants [5,21,22]. In each case the fast neutrons from ^{238}Pu-Be excite silicon by the reaction $^{28}Si(n,p)^{28}Al$, and the 1.78 MeV gamma rays from 2.3 minute ^{28}Al are counted with a NaI(Tl) scintillation spectrometer. The equipment is in routine plant operation.

The U. S. Bureau of Mines are also investigating on-line analysis of iron ore [23] with the objective of monitoring silicon and iron in flowing taconite slurries, and so avoiding the currently used magnetic method for iron that is susceptible to error through insensitivity to non-magnetic iron. A 60 μg ^{252}Cf source and a Ge(Li) spectrometer are used to excite and measure the 3.539 and 4.934 MeV thermal neutron capture gamma rays from $^{29*}Si$ and the 7.631 and 7.645 MeV ones from $^{57*}Fe$. The main problem at present is noise from radiation and acoustic background. Fast fourier analysis of the spectra is being used to reduce the noise contributions.

A second Bureau of Mines group has been testing a prototype coal sulphur meter for several years in pilot plants and for the last two years in a commercial coal preparation plant [24]. The instrument uses an 80 μg ^{252}Cf source and a 6 in. dia. x 7 in. NaI(Tl) spectrometer equipped with specially designed fast pulse processing ($\sim$ 3 x 10^5 cts/sec). The source is centered in a vertically moving 40 in. dia. bed of -1/4 in. coal. The detector counts 5.43 MeV thermal neutron capture gamma rays from the sulphur. Compensation is made for background variations due to changes in ash, iron and moisture contents, and bulk density. Reported accuracies are in the range 0.02 to 0.05% S with an instrument response time of 2 mins.

A dual source technique has been developed for resolving the mutual interference between aluminum and silicon [25] and, together with calcium determination, has been shown feasible for on-stream analysis of cement raw mix [26]. Thermalized neutrons from ^{252}Cf activate aluminum by the reaction $^{27}Al(n,\gamma)^{28}Al$ and calcium by $^{48}Ca(n,\gamma)^{49}Ca$. Fast neutrons from ^{238}Pu-Be activate silicon by $^{28}Si(n,p)^{28}Al$ in a parallel measurement. Using a recirculating, dual slurry-loop system the following measurement precisions are obtained in a 15 minute response time: 0.2% CaO at 44%; 0.1% SiO_2 at 13%; 0.03% Al_2O_3 at 4%. The equipment is offered commercially for on-stream analysis of a wide range of bulk materials [6].

3.1.2. Batch Sample Analyzers

Batch sample analyzers are also available commercially for industrial process control. Several are in operation, or are feasible, for the following analyses: Al, Si in iron ore and bauxite, F in fluorite and glass frit, Si in fluorspar and Na in aluminum silicate [5]. In most cases fast neutron activation is employed using a ^{238}Pu-Be or ^{241}Am-Be source and the emitted gamma rays counted with a NaI(Tl) spectrometer. Samples of 100 to 200 grams are presented manually. Accuracy is 2 to 5% of the amount present in measurement times of 5 to 20 minutes.

A second group is performing field evaluation of a neutron method for determination of cement in cement-soil mixture, a material used in roads and runways. A mobile thermal neutron activation analysis unit housing a 140 μg ^{252}Cf source and a 5 x 5 in. NaI(Tl) detector has undergone successful field trials. It is also being evaluated for determination of cement content of concrete [27].

Perhaps the most sophisticated equipment of this type is the multicomponent plastic concrete analyzer commissioned by the U. S. Army Corps of Engineers [28]. The ultimate strength of concrete depends on the cement and water contents of the plastic concrete mix. Most USA concrete uses a mixture of siliceous and calcareous aggregate, making it difficult to uniquely determine cement content by elemental analysis. The problem is solved by measuring the carbon concentration and so arriving at a value for the carbonate rock content. The nuclear reaction used is $^{12}C(n,n'\gamma)^{12*}C$ and the 4.43 MeV gamma radiation resulting from ^{238}Pu-Be excitation is measured with a 5 in. x 5 in. NaI(Tl) spectrometer. The other key signature elements are Si, H and Ca. The preferred method for measuring Si is to count the prompt 1.78 MeV gamma rays resulting from the same fast neutron excitation used to determine carbon. Hydrogen is measured by thermal neutron excitation using a 180 μg ^{252}Cf source and counting the 2.22 MeV prompt gamma rays. This sample has then undergone thermal neutron activation and yields calcium data from the 3.09 MeV activation gamma rays emitted by ^{49}Ca. The instrumentation consists of two prompt gamma ray cells (one using fast and the others thermal neutrons), an activation gamma ray cell and an electronic and control unit. Four-Kg samples are measured in parallel in the two prompt gamma ray cells and one or both samples subsequently counted in the activation gamma ray cell. In a total analysis time of 10 to 15 minutes, the water, cement, and aggregate contents can be determined to 2 to 5% relative accuracy. The equipment is at present undergoing field trials with the U. S. Army Corps of Engineers.

The same or similar instrumentation can be used for many process analyses of bulk materials not hitherto practicable [6]. These include C, H, O, Al, Si, S, Fe content of coal; C, H, Al, Si, Ca and Fe content of iron ore sinter mix; and N, P, K content of fertilizers.

3.2. X-Ray Instrumentation

One on-stream slurry analyzer has been sold commercially, for determination of iron in taconite concentrates [5]. A ^{238}Pu source is used to excite Fe K X-rays, which are measured with a proportional counter. An on-line density gauge is incorporated and percent iron in the solids is computed automatically. No performance data is available. Such instrumentation has been offered for several years by at least two companies [5,6] and, to our knowledge, has resulted in this single sale.

One "on-stream" and several "off-stream" slurry and solution analyzers based on Si(Li) X-ray spectrometers and radioactive sources have been installed for copper ore process control. As far as we know, only one user [29] and one supplier [30] is significantly involved at present. One such instrument [29] uses a 100 mCi ^{238}Pu source to excite Cu K X-rays in small cake samples filtered from slurry streams.

Immersible probes comprising Si(Li) detectors and radioisotope sources have been investigated with very encouraging preliminary results. Possible microphonics in the detector due to the flow of the stream have been avoided. Work is continuing along these lines [31].

The X- or gamma ray backscatter coal ash monitors long available in Europe are now offered commercially in the USA [32,33] but significant sales have not resulted to date. The high and variable Ca and Fe content of some US coals may inhibit application of these instruments.

ACKNOWLEDGEMENTS

The author is grateful to the numerous people who contributed to this paper by readily submitting to telephone interrogation.

REFERENCES

[1] "Nuclear Techniques and Mineral Resources", IAEA, Vienna, 1969, STI/PUB/198.

[2] "Radioisotope X-Ray Fluorescence Spectrometry", IAEA, Vienna, 1970, STI/DOC/10/115.

[3] "Nuclear Techniques for Mineral Exploration and Exploitation", IAEA, Vienna, 1971, STI/PUB/279.

[4] "Nuclear Techniques in the Basic Metal Industries", IAEA, Vienna, 1973, STI/PUB/314.

[5] Texas Nuclear Corporation, P. O. Box 9267, Austin, Texas 78766, USA, Technical Literature.

[6] Columbia Scientific Industries, P. O. Box 6190, Austin, Texas 78762, USA, Technical Literature.

[7] MATTSON, R. S., COX, J.A., Canadian Mining and Metallurgical Bulletin, April 1973.

[8] NORDVIK, E., GRUBER, B., Norwegian J. Chem., Min. and Metall., 19, 8A (1971).

[9] RHODES, J. R., FURUTA, T., Trans. Instn. Min. Metall., 77 (1968) B162.

[10] MARR, H. E., CAMPBELL, W. J., U. S. Bureau of Mines Report of Investigations RI 7611 (1972).

[11] BERRY, P. F., FURUTA, T., USAEC Reports ORO-3842-1, -2 and -3 (1970).

[12] MITCHELL, T. M., U. S. Federal Highway Administration Report No. FHWA-RD-73-48 (1973) (NTIS PB-224 605).

[13] SWIERKOWSKI, S. P., ARMANTROUT, G. A., WICHNER, R., IEEE Trans. Nucl. Sci. NS-21 (1974) 302.

[14] INAX Instruments Ltd., P. O. Box 6044, Station J, Ottawa K2A 1T1, Canada, Technical Literature.

[15] JAGODA, N., et al., IEEE Trans. Nucl. Sci. NS-21 (1974) 194.

[16] CLARK, B. C., BAIRD, A. K., Earth and Planetary Sci. Lett. 19 (1973) 359.

[17] CLARK, B. C., BAIRD, A. K., Geology, September 1973, 15.

[18] TOULMIN, P., III, et al., Icarus 20 (1973) 153.

[19] PERMAR, P. H., in Proc. Cf-252 Utilization Meeting, San Diego, Calif., CONF-740447-1 (DP-MS-74-27) (1974).

[20] TUTTLE, W. H., WILLIAMS, C. J., PETERSON, G. A., Mining Congress J., Jan. 1972.

[21] CAMPBELL, D. G., FARM, P. D., GLADYZ, C. V., Canadian Min. Metall. Bull., June 1974.

[22] BERRY, P. F., in Proc. 2nd Annual Symp. ISA Mining and Metallurgy Group, Toronto, June 1973.

[23] Californium-252 Progress 16 (Dec. 1973) 24.

[24] STEWART, R. F., et al., U. S. Bureau of Mines Tech. Prog. Rep. TPR 74 (1974), (NTIS PB-228 678).

[25] TAYLOR, M. C., RHODES, J. R., U. S. Patent 3,781,556 (1973).

[26] TAYLOR, M. C., RHODES, J. R., Instrum. Technol. 21 (Feb. 1974) 32.

[27] Californium-252 Progress 16 (Dec. 1973) 30.

[28] TAYLOR, M. C., Analysis Instrum. 12 (1974) 165, Instrument Society of America.

[29] MADDEN, M. L., Eng. Min. J. 174 (Dec. 1973) 84.

[30] Nuclear Equipment Corp., 931 Terminal Way, San Carlos, Calif. 94070, USA, Technical Literature.

[31] LANGHEINRICH, A., Kennecott Copper Corporation, Private Communication.

[32] Sortex Company of N. America, Inc., P. O. Box 160, Lowell, Mich. 49331, USA, Technical Literature on NCB Phase IIIA Ash Monitor (with Iron Compensation).

[33] Superior Electronics, Inc., 1330 Trans Canada Hwy., South, Dorval 740, Quebec, Canada. Technical Literature on "Wedag" Ash Monitor (^{241}Am Backscatter).

REVIEW OF SOME NUCLEAR TECHNIQUES USED IN MINERAL PROCESSING IN THE UNITED KINGDOM

J.F. CAMERON
Nuclear Enterprises Ltd.,
Reading, Berks,
United Kingdom

Abstract

REVIEW OF SOME NUCLEAR TECHNIQUES USED IN MINERAL PROCESSING IN THE UNITED KINGDOM.

In the United Kingdom, instrumentation based on nuclear techniques is extensively used in the mineral processing industries. The major applications are still density and level gauging and those based on portable and laboratory X-ray fluorescence analysers. However, a variety of other methods are being developed and rapidly adopted. New instruments include neutron moisture gauges, conveyor weighers and on-stream analysers.

1. EQUIPMENT FOR CONTINUOUS ANALYSIS OF SLURRIES, PULPS AND POWDERS BY MEANS OF X-RAY FLUORESCENCE

The major factors that have influenced the development of XRF systems for continuous analysis over the past five years have been:

(a) The availability of cheap, small digital computers for dedicated use which has rendered hardwired and analogue computer systems obsolete. The latest minicomputers are robust with fail-safe and automatic restart capability and are eminently suitable for real-time industrial applications.

(b) The improvements in rugged, non-microphonic, reliable high-resolution semiconductor detectors suitable for continuous industrial use.

(c) The development of miniature air-cooled end-window X-ray tubes for excitation sources of high intensity. When semiconductor detectors are used, these tubes permit counting times to be reduced to acceptable limits.

Analysis of one or two elements

There are many applications where the analysis of one or two elements is adequate for control, and for this purpose a relatively cheap system is available based on radioisotope sources, proportional and/or scintillation counters and a minicomputer [1, 2]. One system of this type has been in continuous operation for over two years monitoring Sn tailings [3, 4].

Satisfactory results are now being obtained in the continuous determination of Ca in cement raw meal with a ^{55}Fe source, a sealed proportional counter and a special powder presenter which gives a constant mass per unit are [5]. Early set-backs were experienced mainly as a result of environmental factors such as high temperature and dust, but satisfactory solutions have been evolved.

Multi-element analysers

Systems incorporating a semiconductor detector, a minicomputer and either an X-ray tube or a radioisotope source are now available commercially for multi-element analysis [1, 2, 6]. In general, the use of the radioisotope source is restricted to cases that require an excitation energy above 25 keV. One such installation is for the analysis of Y and rare earths in solution when ^{57}Co and ^{109}Cd sources are used.

For slurry analysis, essential ancillary equipment is a primary sampler, a filter, a secondary sampler, de-airing and constant head devices, a density gauge (gamma-ray absorption system) and a sample presenter. The sample presenter must be designed to maintain sufficient turbulence to present a representative sample and yet not cause excessive window abrasion.

A special analyser for determining Al, Si, Fe and Ca in cement has been developed and is now being installed for on-line trials [6]. Sample presentation is by means of a specially developed belt unit which gives high positional accuracy and near constant bulk density. A helium path is provided between the X-ray tube and the Si(Li) detector. Two sequential measurements are made: one with the tube at 1 mA and 3.5 kV (to avoid exciting Ca) to determine Al and Si, and the second at 0.01 mA and 10 kV to determine Ca and Fe. For 29 different samples, each run 8 times, the relative standard deviations for Si, Al, Fe and Ca at mean concentrations of 15, 3, 1.5 and 45% were 0.7, 1.8, 0.7 and 0.2% respectively.

A windowless sample presenter has been developed in which the sample is pumped in a jet from a silicon nitride nozzle [7]. This system permits measurements of low Z elements as well as measurements on highly corrosive or abrasive fluids. An analyser for coarse solids is being developed. It incorporates an X-ray tube or a radioisotope source, an energy dispersive spectrometer and a special sample presentation system. Numerical methods and computer programs are being developed to give more accurate quantitative results.

2. IN-STREAM ANALYSER

An XRF probe that can be immersed in any sample stream to determine the concentration of a single element has been developed and used in field trials [8, 9]. A commercial system is under development which will incorporate a real-time control and data processing unit based on a minicomputer. The probe incorporates a gamma transmission gauge to determine pulp density, safety interlocks and a calibration check facility. Field trials have been held in a copper mine and the feasibility of measurements on tin and zinc tailings has also been proven.

For satisfactory operation of such a probe it must "see" a representative sample. Consequently, practical problems that are extremely difficult or even impossible to overcome often arise from stratification and aeration of the sample.

Although a single element system of this type is cheaper than a simple parallel stream system, the latter is capable of handling automatically six or more sample streams in sequence. It also provides a superior performance since errors due to stratification, aeration and deposition on windows are avoided by the sampling and presentation system.

A new low-cost immersible XRF probe with a low-power XRF tube or a radioisotope source, a solid-state detector and a microprocessor for data handling is being developed.

3. ON-STREAM PARTICLE SIZE ANALYSIS

A system based on the Mintek/R.S.M. slurry sizer [10] has been working for over a year in an iron-ore mill in Canada and a second one has recently been installed in a copper mine in Norway. Results indicate that an accuracy of about 2% (1σ) can be achieved for material passing 200 mesh on-stream.

4. ON-STREAM NEUTRON ACTIVATION ANALYSIS

The recirculation neutron activation technique [11] has been used in three systems in the USA and Canada for the determination of silica in iron ores.

5. ON-STREAM NEUTRON ABSORPTION ANALYSIS OF SOLUTIONS

Four systems have been supplied which monitor continuously the boron content in nuclear fuel-element cooling ponds in the concentration range 750 - 2 000 ppm, and a further two systems are under construction. The sample is flowed through an annular chamber and a 100-mCi ^{241}Am source and a ^{3}He neutron counter are mounted in the centre of the annulus.

6. DENSITY AND LEVEL GAUGES

Nucleonic density and level gauges are being used on an ever-increasing scale in the mineral processing industries. The nucleonic density gauge, which is mounted on the outside of a pipe carrying the mineral slurry, is easily installed and serviced. It gives a measurement integrated over a large representative portion of the pipe cross-sectional area, does not interfere with the flow and is unaffected by the abrasive nature of the product.

The biggest user of nucleonic level gauges (both gamma switches and continuous level gauges) is the minerals processing industry because the gauges are mounted on the outside of the process vessels. This means that they are unaffected by the bumps from lumps of ore, by the corrosion or abrasion caused in the processing, e.g. by processed fluids, by the dusty or wet environment, and by the extreme temperatures.

7. NUCLEONIC BELT WEIGHERS

In mineral processing the correct rate of feed of ore throughout the processing plant (crushers, grinders, flotation cells etc.) is controlled by belt weighers. In this type of environment the nucleonic belt weigher [12, 13] is proving far superior to the load cell weighers for reliability, high availability, ease of calibration check and computer compatibility. It is also easier to install in existing conveyors without modification to the conveyor, and can be used on inclined conveyors and conveyors of virtually all types (bucket, screw etc.). Likewise, it can be used near the point of feed which is extremely important for tight feed-back control since the velocity-distance lag is reduced to a minimum. Up to 70 or 80 nucleonic weighers may be installed on a single large mining and processing plant.

8. ON-LINE NEUTRON MOISTURE GAUGES

Neutron moisture gauges [14, 15] are ideal for measurements on materials such as ores and furnace coke since they give a non-contact measurement integrated over a large representative volume of material. So far most experience has been gained on furnace coke. Measurement on coke is essential since the ratio of coke to the other constituents charged is an important parameter to control, and the coke moisture content can vary from 0 to 20% by weight. The moisture gauge measuring head is generally mounted on the wall of a hopper containing the coke and the signal from the electronics unit can be used for manual or automatic [15] control of the dry weight charging.

9. PORTABLE AND LABORATORY XRF ANALYSERS

Although small XRF analysers based on radioisotope sources and balanced filters have been available for about 12 years, it is only now that they are becoming recognized as a very useful, reliable, low-cost analyser [16]. Many of the earliest instruments were not sufficiently robust and reliable, but experience over the years has resulted in design suited to this type of application. Errors resulting mainly from ore heterogeneity and matrix variations led to many early users becoming frustrated, but simple methods of allowing for these variations have been evolved. Thus, provided that careful instructions are given to the operators (who are often unskilled) and that the results are properly interpreted and corrected (here the advent of cheap pocket electronic calculators has been a great help), excellent analyses can be achieved. Several major ore bodies have been discovered with these instruments and they are also playing an important role in mine control.

REFERENCES

[1] STARNES, P.E., The Mintek multielement on-stream analyser for mineral dressing plants, Mintek/Topic 4, Cartner Group Ltd., Slough, U.K.

[2] STARNES, P.E., Comparison between the Mintek multielement non-dispersive on-stream slurry analyser and conventional dispersive on-stream analysers, Mintek/Topic 5, Cartner Group Ltd., Slough, U.K.

[3] STARNES, P.E., Essential technical facts relating to the Mintek system for on-stream determination of tin at Geevor Mine, Pendeen, Cornwall, Mintek/Topic 2/3, Cartner Group Ltd., Slough, U.K.

[4] EDITOR, Continuous tin analysis, Tin International (Nov. 1973).

[5] STARNES, P.E., Private communication.

[6] CARR-BRION, K.G., On-stream energy dispersive X-ray analysers, X-ray Spectrom. 2 (1973) 63.

[7] CARR-BRION, K.G., U.K. Patent 46105/71.

[8] WILLIAMS, A.W., CARR-BRION, K.G., "The use of an in-stream head for the X-ray fluorescence analysis of slurries," Proc. Conf. on Industrial Measurement and Control by Radiation Techniques, Institute of Electrical Engineers, London (1972) 94.

[9] CARR-BRION, K.G., WILLIAMS, A.W., In-stream determination of copper — a new concept in control analysis, Mining Magazine 127 No. 4 (1972).
[10] STARNES, P.E., Monitoring wet mill performance with the Mintek/R.S.M. slurry sizer, Mintek/Topic 2/2, Cartner Group Ltd., Slough, U.K.
[11] STARNES, P.E., Improvements in activation analysis, U.K. Patent 1,124,992 (1966).
[12] BOYCE, I.S., CAMERON, J.F., PIPER, D., "Nucleonic conveyor-belt weighers for the basic metals industry", Nuclear Techniques in the Basic Metal Industries (Proc. Symp. Helsinki, 1972), IAEA, Vienna (1973) 155.
[13] CAMERON, J.F., KING, M., BRISTOW, J., Nucleonic conveyor weighers SIMAC 74, Inst. of Measurement and Control (1974).
[14] PAPEZ, K., CAMERON, J.F., MACHAJ, B., "On-line neutron moisture gauges in the steel industry", Nuclear Techniques in the Basic Metal Industries (Proc. Symp. Helsinki, 1972), IAEA, Vienna (1973) 183.
[15] CAMERON, J.F., Computerized control of blast furnace coke-weighing operations using neutron moisture gauges, Eurisotop Office Information Booklet 91, Brussels (1974).
[16] PAPEZ, K., CAMERON, J.F., "Applications of nondispersive X-ray fluorescence analysers in the basic metals industry", Nuclear Techniques in the Basic Metal Industries (Proc. Symp. Helsinki, 1972), IAEA, Vienna (1973) 21.

ON-LINE DETERMINATION OF THE IRON CONTENT OF ORES, ORE PRODUCTS AND WASTES BY MEANS OF NEUTRON CAPTURE GAMMA RADIATION MEASUREMENT

K. LJUNGGREN, R. CHRISTELL
Isotope Techniques Laboratory,
Stockholm, Sweden

Abstract

ON-LINE DETERMINATION OF THE IRON CONTENT OF ORES, ORE PRODUCTS AND WASTES BY MEANS OF NEUTRON CAPTURE GAMMA RADIATION MEASUREMENT.

A method for direct continuous determination of iron in ores, ore products and wastes has been developed. The method is based on the measurement of the 7.64-MeV gamma radiation emitted on capture of thermal neutrons in the nuclei of the iron isotope ^{56}Fe. A prototype instrument, which contains a ceramic 5-Ci ^{238}Pu-Be source, is being evaluated on a production line for crushed hematite ore by ITL in collaboration with the mining company LKAB. The method permits measurements to be made with a precision of 1% (CV) or better over the iron concentration range of interest. The method is described and the tests performed in the laboratory and in the plant are reported.

1. INTRODUCTION

A method for direct continuous determination of iron in ores, ore products and wastes has been developed and a prototype instrument has been designed and installed on a process stream. The iron determination is based on the measurement and recording of the high-energy gamma radiation (7.64 MeV) emitted on capture of thermal neutrons in the nuclei of the iron isotope ^{56}Fe, as indicated in Fig.1 [1]. The method is being evaluated by ITL in collaboration with the mining company LKAB.

The prototype stage was preceded by laboratory and plant measurements on stationary as well as on moving products. The method is considered very promising for process control as it allows a continuous selective determination of iron in a large representative sample to be made directly on the process stream.

The neutron source used for most of the measurements was a 3-Ci ^{241}Am-Be source, emitting 7.5×10^6 n/s, placed in an aluminium tank containing water as a moderator. For the prototype installation, a ceramic 5-Ci ^{238}Pu-Be source, emitting 4.2×10^6 n/s was used.

The gamma radiation is recorded spectrometrically in selected energy channels by means of a 50 mm $\phi \times$ 50 mm NaI(Tl) scintillation detector. Automatic spectrum stabilization has been used in part of the tests and is used in the prototype equipment. A schematic view of the equipment is given in Fig.2.

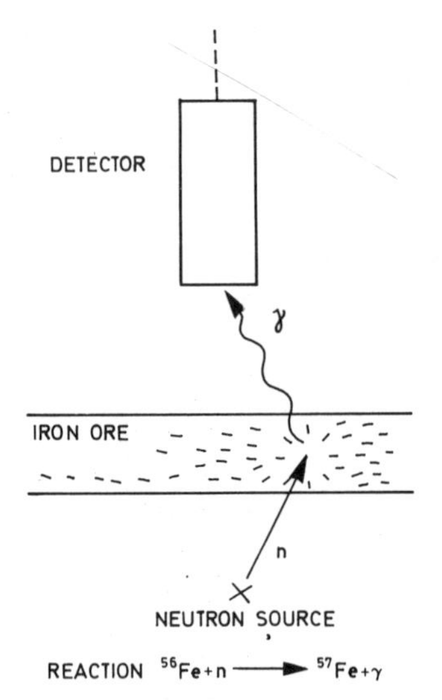

FIG.1. Principle for the determination of the iron content of ores, etc. by measurement of neutron capture gamma radiation. The intensity of the characteristic high-energy capture gamma radiation (7.64 MeV) is a measure of the iron content.

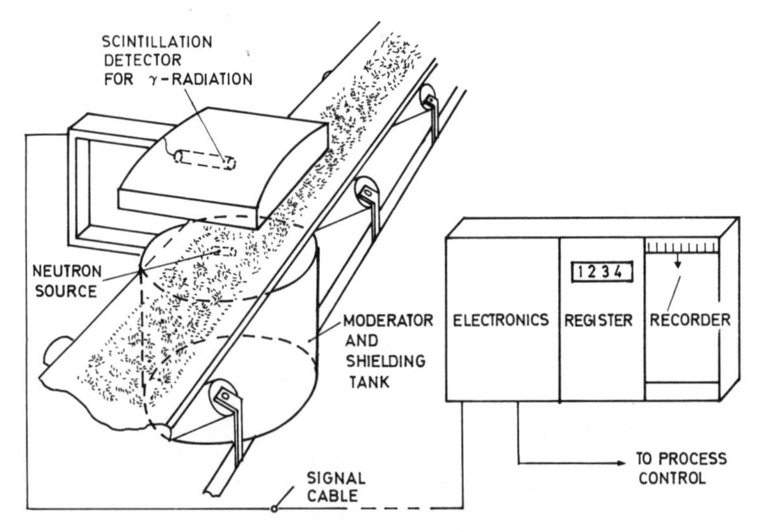

FIG.2. Iron content analyser.

2. LABORATORY TESTS

The influence of the measured signal of packing, material thickness, grain size, moisture content and vertical and horizontal position of the belt load in relation to the source/detector set-up has been investigated. The iron content, as determined by standard wet chemical analysis of sampled material, ranged from 25 to 70% weight. The results show that the iron content of a stream of crushed ore and ore products can be determined with a precision of within 2% (CV), provided that the geometry of the stream at the detector is properly profiled and only moderate variations in the physical structure (grain size, packing etc.) of the stream occur.

From these experimental investigations, in which the different parameters were varied separately as far as possible, the following (apparent) deviations in the measured iron content could be deduced.

Conditions (stationary ore samples)	Deviation, % (relative)
Repeated measurements on one and the same sample	± 0.5 - 1.0 (CV)
Variation in packing degree: loose-hard	Max. ± 4.4
Variation in moisture (normal moisture content is 1 - 2% H_2O by weight; maximum variation 1%)	+ 1.3 for an increase of 1% H_2O by weight

3. PLANT TESTS

In 1971, a pilot set-up for continuous determination of iron directly on belt-transported ore was arranged and run at the LKAB works at Malmberget (grain size - 30 mm). Repeated measurements on the same material showed good reproducibility (CV $\leq$ 1%). However, measured values and values obtained by chemical analysis occasionally differed by as much as several per cent. It is believed that these discrepancies are mainly due to non-representative sampling. To verify this hypothesis, accurately weighed mixtures of iron ore and attle rock were measured. The result of these measurements is presented in Fig. 3, which shows the relationship between measured gamma radiation intensity and iron content. The deviation from a straight line, as determined by the method of least squares, corresponds to a standard deviation of 2% Fe at a concentration of 50% Fe, which was considered satisfactory.

A new series of test runs were made on crushed hematite ore in 1972. Materials in two distinctly different concentration ranges were measured. The low concentration range (35.3 - 49.1% Fe) consisted of wastes and mixtures of ore concentrate and wastes, while the high concentration range (62.9 - 69.0% Fe) covered various mixtures of concentrates. The gamma-radiation intensities (counts in 2 min) have been plotted against concentration averages obtained by chemical analysis of sampled material, see Figs 4 and 5. The regression lines are based on least-squares calculations. In addition, the 95% confidence limits for these lines have been included in the diagrams.

Assuming that the values obtained by chemical analysis are correct and using the regression line as a calibration line, the 2σ-error at mid-range concentrations can be estimated at 0.7% Fe for both low and high iron content.

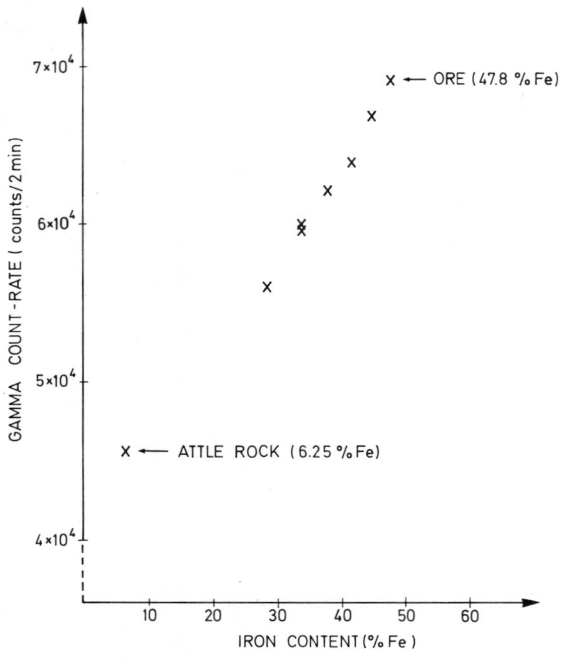

FIG.3. Gamma-ray count-rate plotted against iron content of ore/attle rock mixtures.

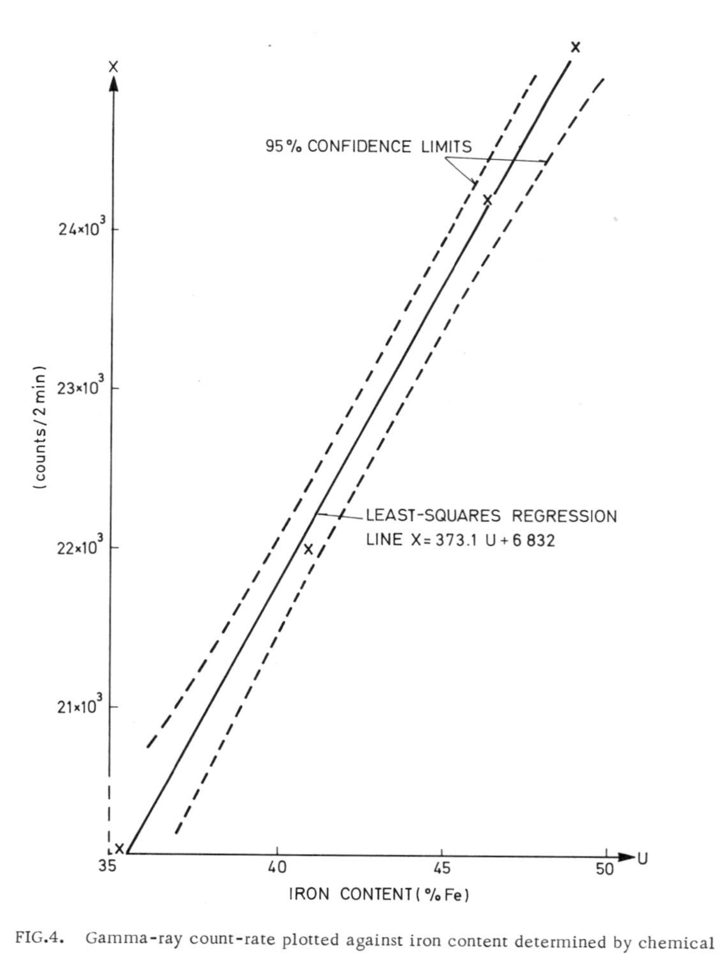

FIG.4. Gamma-ray count-rate plotted against iron content determined by chemical analysis. (Low iron content.)

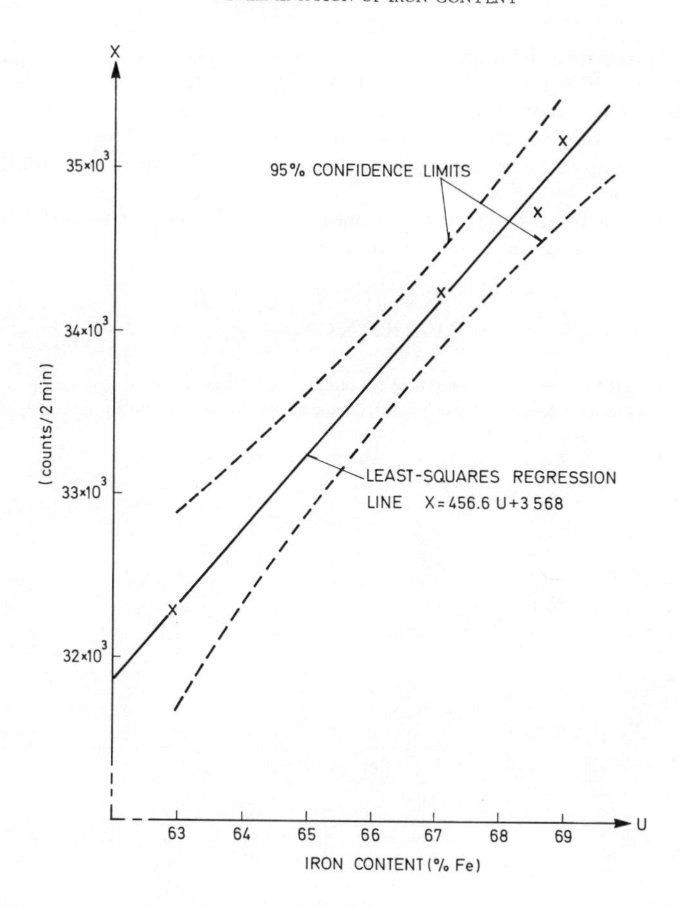

FIG.5. Gamma-ray count-rate plotted against iron content determined by chemical analysis. (High iron content.)

In practice, the reference values for the iron concentration are certainly not free from errors, and a large number of comparative measurements will have to be made if an accuracy of the order mentioned is to be obtained. This is a general difficulty when a new method of measurement depends on an established method of limited precision and/or accuracy for calibration.

4. PRESENT SITUATION

A prototype instrument is at present being tested on a production line for crushed hematite ore [2]. Control engineering in hematite ore processing by means of dry methods is particularly interesting because special concentrating techniques are used and the choice of measuring methods is very limited as magnetic methods fail. A long-term test of the prototype under plant operating conditions has just been started and will continue for at least 6 months.

A prerequisite to the installation of the prototype instrument, imposed by the radiation safety regulations, was an analysis of the safety aspects in heavy industrial operation of the use of radioactive neutron sources, based on beryllium combined with an alpha radiation emitter. The safety features found necessary have been incorporated in the system as described in the patent application [2].

The work has been financially supported by the Swedish Board for Technical Development and the LKAB company.

REFERENCES

[1] CHRISTELL, R., LJUNGGREN, K., "Analysis of ore samples using gamma-rays emitted under irradiation with a low-level neutron source," Radiochemical Methods of Analysis (Proc. Symp. Salzburg 1964) 1, IAEA, Vienna (1965) 263.

[2] CHRISTELL, R., KOSKI, K., LJUNGGREN, K., Swedish Pat. Appl. 74-04824-0.

PHYSICAL ASPECTS OF THE APPLICATION OF NUCLEAR TECHNIQUES IN MINING EXPLOITATION

K. PRZEWŁOCKI
International Atomic Energy Agency,
Vienna, Austria

Abstract

PHYSICAL ASPECTS OF THE APPLICATION OF NUCLEAR TECHNIQUES IN MINING EXPLOITATION.

The paper discusses the results of investigations made to date in mines on the physical parameters of the continuous media by means of measurement methods derived from nuclear physics. As an example, experimental global and local characteristics of movement of the inhomogeneous fluids in mining pipe lines expressed as a function of the mean density, local distribution of concentration of the solids, their acceleration, and intergranular shipping velocities, are presented. The experimental determination of the coefficient of compressibility of the deposited back fill is described. Measurements can be performed by means of the commercial soil density and moisture gauges with small adjustments. The latter examples are taken from mining hydrology. It was shown that underground waters in mines indicate high variations in the stable isotope composition (^2H, ^{18}O). In some cases this can be used to distinguish between static and dynamic water resources.

I. INTRODUCTION

Monitoring nuclear radiation in uranium mining was started over 40 years ago. It was used to facilitate exploitation of the ore body and to protect miners against excess irradiation.

For the last 15 years, nuclear techniques, in a more general sense, have been successfully implemented in mining technology of non-radioactive mineral raw materials. More industrial and physical parameters can now be measured and controlled by means of radioisotope gauges. The application of the following radioisotope gauges and methods based on nuclear techniques are described in the literature:

1. Automation of underground operations by means of radioisotope relays [2].
2. Control of wear of stowing pipe lines by means of portable radioisotope scattering gauges [2].
3. Automatic driving of the coal combine by means of a radioisotope sensor [3].
4. Control of the stowing mixture consistency by means of a radioisotope density gauge [4].
5. On-line determination of coal moisture by means of a neutron gauge [5].
6. Radioisotope conveyor weights [6].
7. Radioisotope classifier of the granulometric composition of coal in mining cars [7].
8. Industrial analysis of polymetallic ores by means of photoactivation [8, 9].
9. Fast analysis of the ore grade at the mining face [10].

10. Determination of potassium in potassium mines and processing factories [11].
11. Analysis of rocks and ores by means of neutron activation [12,13].
12. Assay of the ash content of coal [14].
13. Identification of industrial systems in mineral processing factories by means of radioactive tracers [15].
14. Determination of the aluminium content during processing of bauxites [11].
15. Control of ore processing by means of an on-line nuclear analysis system [16].

Most of these examples require special, sometimes quite complex, equipment specially adjusted to work under mining conditions. Commercial companies are trying to manufacture equipment to meet these difficult requirements (100% relative humidity of the air, dust, cold, spark safety, low-qualified operators, etc.). Such an activity is, however, very costly and labour consuming and this is possibly the reason why there are only a few tens of commercial producers of nuclear laboratory equipment and only a few producing nuclear industrial gauges. Often this second activity is financially supported by the national Atomic Energy Authorities. However, one may expect that some of the applications mentioned will be further implemented in mining. The extent of their implementation will depend, on the one hand, on the cost of manufacturing the equipment and on the technical level of the processing and mining industry and, on the other hand, on the value of useful information for the technological process.

This paper considers the more basic aspects of the possible application of nuclear techniques in mining that are not yet covered by commercial activity and are rather due to the environment of the mining activity. For example, the miner, during his work, is in close contact with rocks (solids), natural and technological waters (fluids) and the mining atmosphere (gaseous phase).

When considering this environment, the phenomenological approach has, until now, prevailed; the mechanics of continuous media are a primary tool to describe the behaviour of the mining environment. This approach often does not guarantee an appropriate accuracy of the description. Moreover, due to the lack of appropriate methods for measuring the parameters of the continuous media, the theoretical results often cannot be reliably verified experimentally.

Owing to these difficulties, engineers worked out practical approaches which partially solved the technical problems, but did not contribute enough to the recognition of the physical mechanism of the investigated process. As a result, the literature contains a number of empirically established correlations of the type of "the best fit formula". Their practical usefulness is, in most cases, limited to certain boundary conditions.

This can be illustrated by an example from hydrotransportation which is largely used in the mining industry for transporting the hydraulic back fill, slurries and concentrates in the ore processing factories and in waste disposal, etc.

The problem of hydrotransportation is not only restricted to mining. Economical calculations show that the transportation of big masses of loose materials in the form of hydromixtures moving in steel pipe lines is cheaper

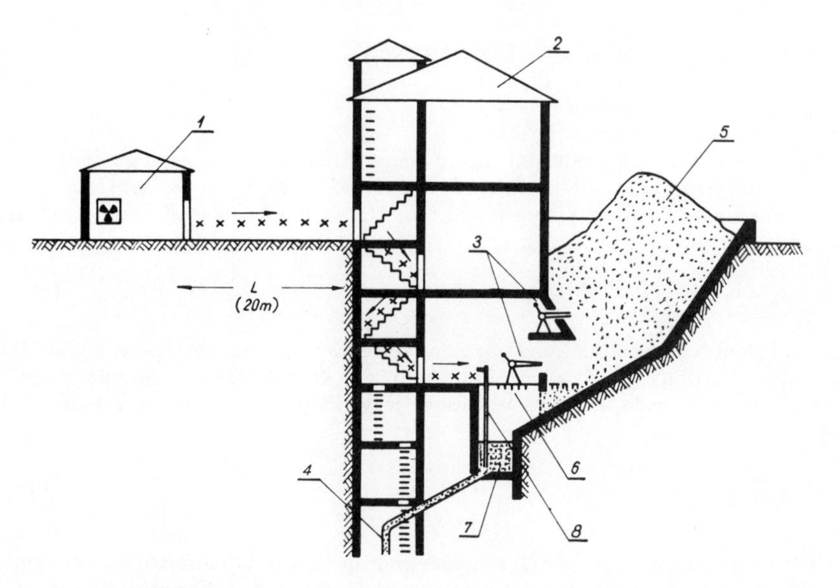

FIG.1. Washery building; positioning of the measuring devices and the installation equipment. 1. Hut; 2. Washery building; 3. Water guns; 4. Pipe-line; 5. Sand; 6. Grating; 7. Retention reservoir; 8. Throw-in injector.

and economically more justified than transportation by wheels, at least for short and medium distances.

Hydraulic back fill is a special case where the movement of the hydromixture is reinforced by gravitational forces. It is used for filling underground cavities, caused through mining exploitation, with sand or crushed rocks, transported through pipe lines to the mining face. After deposition, the back fill protects buildings on the surface against subsidence, and the mining roof against sudden collapse. Figure 1 shows the technological process of the back filling. Hydromonitors mix sand with water and the hydromixture enters the inlet of the pipe line at the surface. A pipe line a few kilometres in total length conducts the hydromixture to the mining face and discharges it there. After deposition of solids, the filtrated off-water is pumped to the day surface and used again for the same technological process.

However, in all cases where hydrotransportation operates, the horizontal plane movement of the inhomogeneous fluid in the pipe line must be reinforced by means of electrical power. Therefore, it is logical that the main criterion for the construction of industrial installations for hydrotransportation must be the optimization of all parameters of the technological process to minimize the required power consumption.

The following procedure was worked out to meet these requirements. If pure water is transported by the pipe line and the flow is turbulent, one can determine the pressure head, I, according to Darcy's law, using the following formula:

$$I_o = \frac{\lambda}{2\,g\,D}\,v^2 \qquad (1)$$

where λ = linear coefficient of resistance
D = diameter of the pipe line
g = acceleration of gravity
v = mean velocity of flow

If some solids are suspended in the moving fluid, the total energy losses for transportation must be higher as additional work must be performed for the suspension and friction of solids. Head losses in this case must be expressed by

$$I = I_0 - \Delta I \tag{2}$$

where ΔI denotes the additional pressure head required for the transportation of the solid phase. Experimentally it was found that the second term on the right-hand side of Eq. (2) is inversely proportional to the mean velocity of flow

$$I = I_0 + \frac{B}{v} \tag{3}$$

where B is an empirical coefficient depending on the parameters characterizing the pipe line and on those of the hydromixture. Formula (3) is widely used by designers of industrial installations for hydrotransportation of solids [17].

Different authors have presented different formulae to describe the coefficient B. These relationships were determined theoretically or experimentally using special testing installations. Table I presents some proposals published in the literature.

A comparison of the formulae in Table I reveals that different parameters occur in different formulae. This gives rise to the following questions: What is the physical meaning of parameter B? In which intervals of the flow parameters are these formulae valid? How accurately do these formulae determine B? Does the choice of the parameters depend on the intuition of the authors or are there more general criteria for the proper selection of these parameters? To answer these questions, more information on the physical mechanisms of the flow of inhomogeneous mixtures in pipe lines is required.

About 20 years ago it was recognized that the theoretical approach alone is not sufficient to properly describe the movement of hydromixtures because of the complexity of the physical phenomena occurring in the flow of the highly condensed hydromixtures [24, 25]. The only approach from which new information can be expected is the experimental one.

During recent years it was shown that the measuring methods based on the application of nuclear techniques offer a high accuracy in the measurement of characteristics of hydrotransportation under conditions of undisturbed flow. In most cases, this accuracy is higher than that offered by conventional measuring methods. Thus, the use of methods based on nuclear techniques open new possibilities for further investigation of physical aspects of flow of inhomogeneous fluids in industrial pipe lines.

In mining, there are more problems of this type concerning the movevent of rocks, water and gases (air, methane, carbon dioxide, nitric oxides, etc.) which could be approached by means of measuring methods based on nuclear techniques. This opinion can be illustrated by the following three

TABLE I. PROPOSALS FOR THE DETERMINATION OF COEFFICIENT B
(Established by K. Korbel [17])

Author of the formula	Formula	Ref.
W.E. Wilson	$B = K\, c_m\, w$	[18]
W.W. Kotulski	$B = K' \dfrac{\gamma_m - \gamma_f}{\gamma_m} c_m w$	[19]
R.C. Worster	$B = 120\, c_v\, \lambda\, (gD)^{\frac{1}{2}} \left(\dfrac{\gamma_M - \gamma_f}{\gamma_f}\right)^{\frac{3}{2}}$	[20]
R. Durand	$B = 40.5\, c_v\, \lambda\, (gD)^{\frac{1}{2}} \left(\dfrac{1}{c_x}\right)^{\frac{3}{4}} \left(\dfrac{\gamma_M - \gamma_f}{\gamma_f}\right)^{\frac{3}{2}}$	[21]
A.E. Ivanov	$B = w \left(\dfrac{\gamma_m - \gamma_f}{}\right) \dfrac{1}{\eta}$	[22]
D.M. Newitt	$B = 550\, c_v\, \lambda \left(\dfrac{\gamma_M - \gamma_f}{\gamma_f}\right) w$	[23]

In these formulae the following notations are used:

K, K' = empirical coefficients depending on the concentration of solids
c_m = mass concentration of solids
c_v = volume concentration of solids
γ_m = mean specific gravity of the mixture
γ_f = specific gravity of the transporting fluid
γ_M = specific gravity of solids
w = velocity of free fall of solids in the transporting fluid
D = diameter of the pipe line
c_x = a coefficient depending on Reynolds number
η = viscosity

particular examples in mining where nuclear measuring methods were successfully tested:

1. hydrotransportation by means of mining pipe lines;
2. determination of parameters of the deposited back fill;
3. determination of the origin of mining waters.

A more detailed discussion of these problems is given in Section II.

II. MEASUREMENTS OF THE FLOW PARAMETERS OF INHOMOGENEOUS FLUIDS IN INDUSTRIAL PIPE LINES

In connection with the example of the hydrotransportation problems discussed in Section I, a proposal was made to measure such parameters as spatial distribution of concentration of solids, acceleration of particular grains of solids and slip velocities by means of the methods based on nuclear techniques.

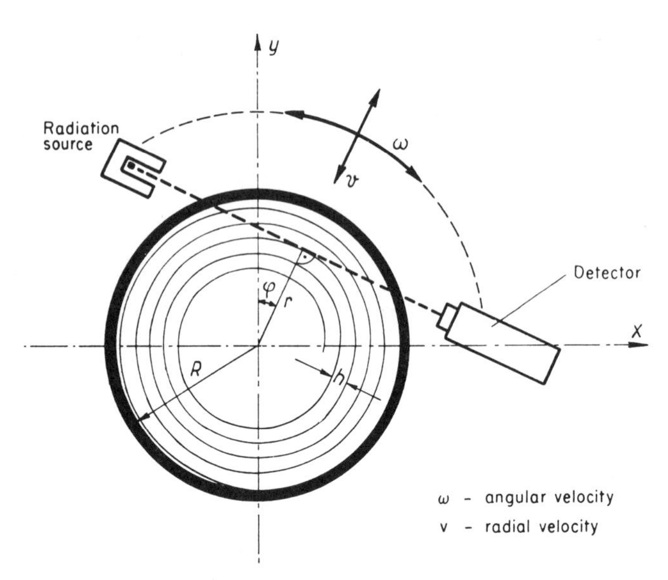

FIG.2. Scanning system for a spatial density distribution.

II. 1. Spatial distribution of solids concentration in the pipe line

The measurements are based on the assumption that if the mean density of the hydromixture at the inlet of the pipe line is kept constant, then the distribution of the concentration of the solids is kept constant too. Figure 2 illustrates a measurement concept. The pencil-narrow beam of gamma radiation scans the cross-section of the pipe line. The whole cycle takes about 2 min [26]. As a result, a two-dimensional concentration pattern can be obtained with an accuracy of measurement of about 5% [27].

Two concentration patterns, taken at the test installation with constant mean velocity for two different volume concentrations, are presented in Figs 3 and 4.

On the right side of Figs 3 and 4, central concentration profiles are shown; the ordinates represent heights above the bottom of pipe Y normalized to its diameter D. On Fig.4 one can see a "core" of concentration well above the bottom of the pipe. The results of a closer investigation of this effect are presented in Fig.5 which shows a family of central vertical profiles of concentration of polyfractional sand, with the mean grain diameter d_{50} = 0.4 mm, in a stream of water [29]. One can observe that, for the low concentration, the amount of suspended solids increases approximately exponentially, with a decrease of height. For higher concentrations another force appears (probably due to a change of the velocity profile) which is responsible for lifting a highly condensed "core". This is a favourable phenomenon as it tends to decrease the wear of the pipe line through friction. Also, it offers lower energy losses. It has, therefore, to be recommended in practice to keep the lowest mean exploitation density of hydromixture on the inlet to the pipe line above some critical value, when the effect of lifting the dense core within the pipe line appears.

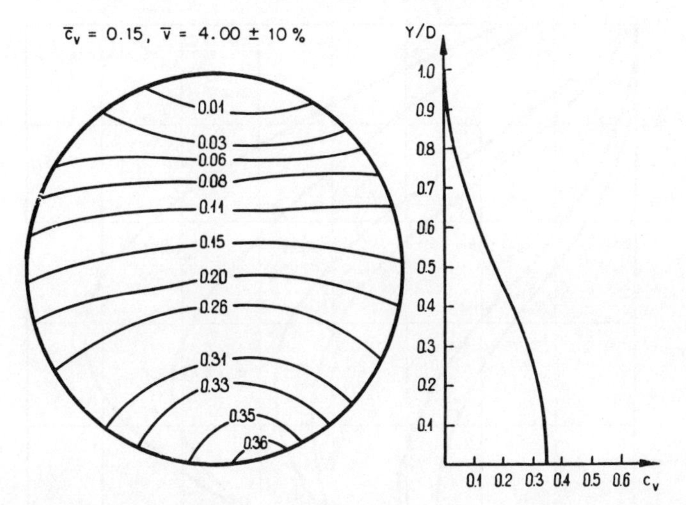

FIG.3. Distribution and profiles of volumetric concentration in the cross-section of the pipe line for the mean concentration $\bar{c}_V$ = 0.15 and mean velocity $\bar{v}$ = 4 m/s. Diameter of the pipe = 175 mm [28].

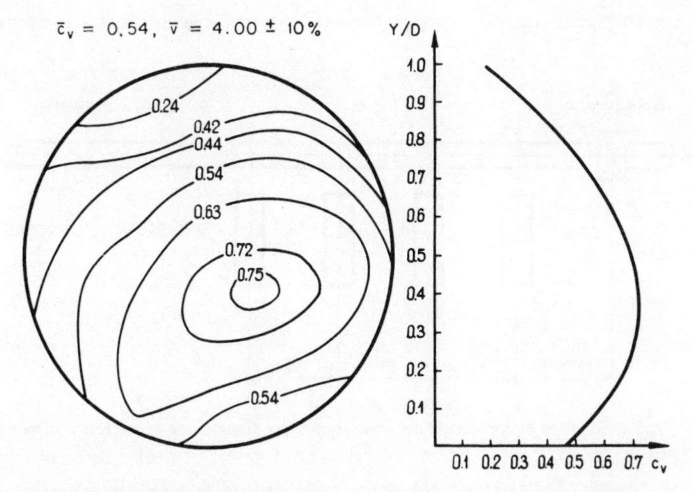

FIG.4. Distribution and profiles of volumetric concentration in the cross-section of the pipe line for the mean concentration $\bar{c}_V$ = 0.54 and mean velocity $\bar{v}$ = 4 m/s. Diameter of the pipe = 175 mm [28].

II. 2. Slip velocities

All industrial density gauges measure so-called "cross-section" mean density. However, existing theories of flow of the inhomogeneous fluids foresee different flow velocities for both phases due to the resistance of the suspended solids to the transporting fluid. If this is true, then the "transport" density of the hydromixture is not the same as the "cross-section" density. What is then the difference? How big are slip velocities? What is the mechanism of acceleration of solids?

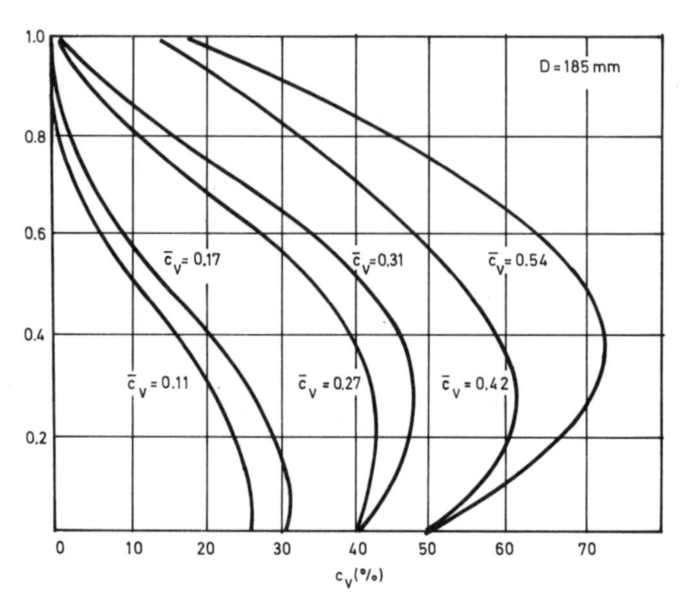

FIG.5. Central vertical profiles of concentration in the industrial pipe line for hydrotransportation of solids [29].

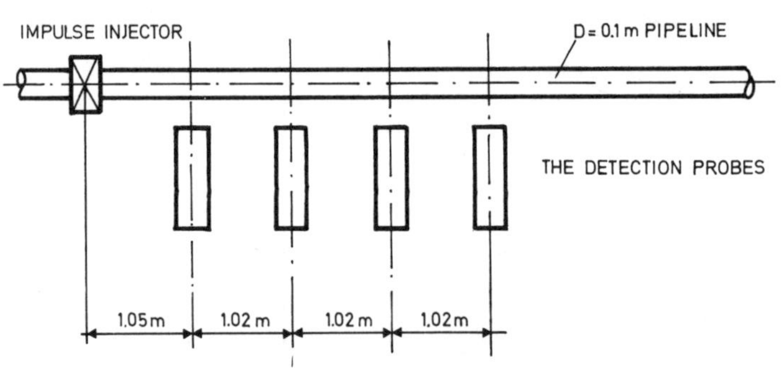

FIG.6. The distribution of detectors during tracer measurements of the acceleration distance [29].

It is possible to design experiments to measure these parameters in the stream of the highly condensed hydromixture by means of radioactive tracers. Moreover, this is probably the only way to measure these parameters.

Figure 6 illustrates the idea of an installation to measure the acceleration of solid particles injected into the pipe line with the flowing hydromixture by means of the special fast injector.

For scintillation, well-collimated probes connected parallel to the multiscaler recorded the velocity wave of the radioactively labelled sand grains. The result of one experiment is presented in Fig.7 [30]. One can notice the effects of inertia on the solids, especially at a range of low velocities of the hydromixtures. It was proved, by means of similar experiments performed for different mean concentrations of hydromixture, that in all cases the accelerating distance does not exceed 15 pipe diameters.

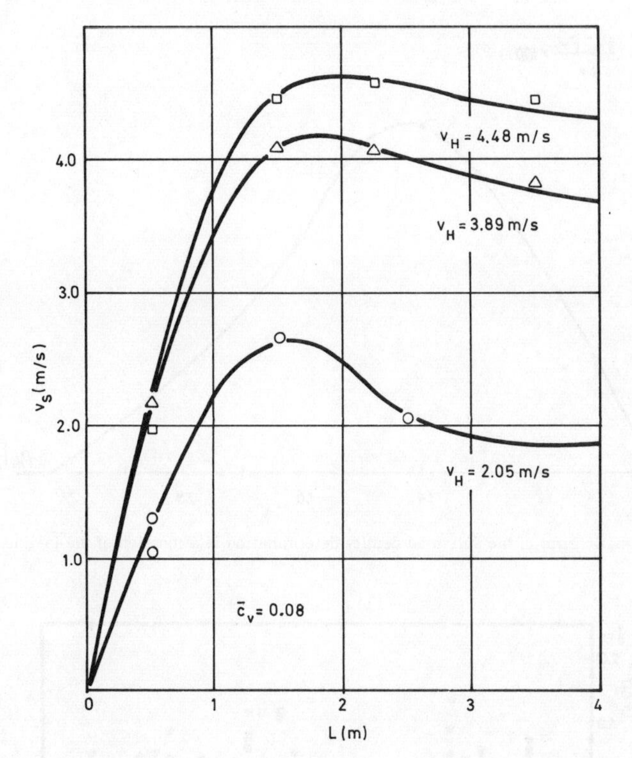

FIG.7. The acceleration of solid particles for the average concentration $\bar{c}_V = 0.08$.

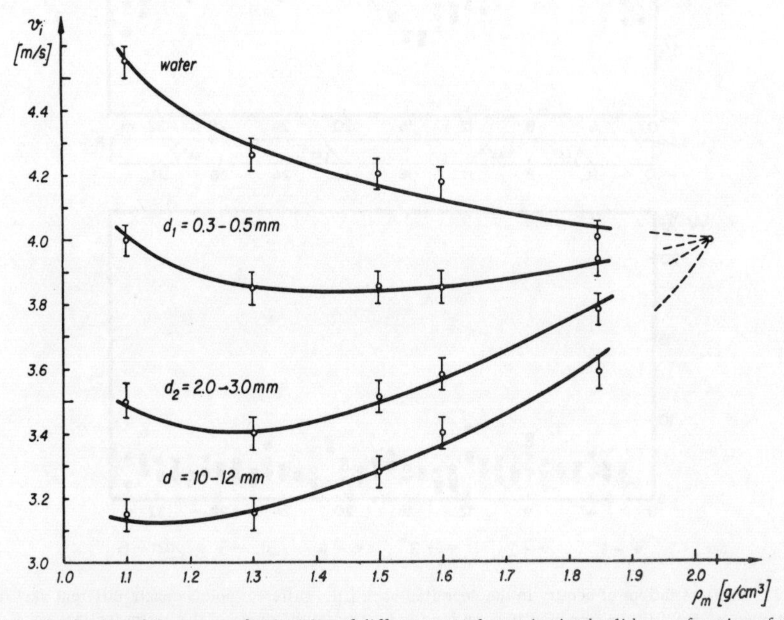

FIG.8. The family of the curve of velocities of different granulometric sized solids as a function of the mean density of the hydromixture [31].

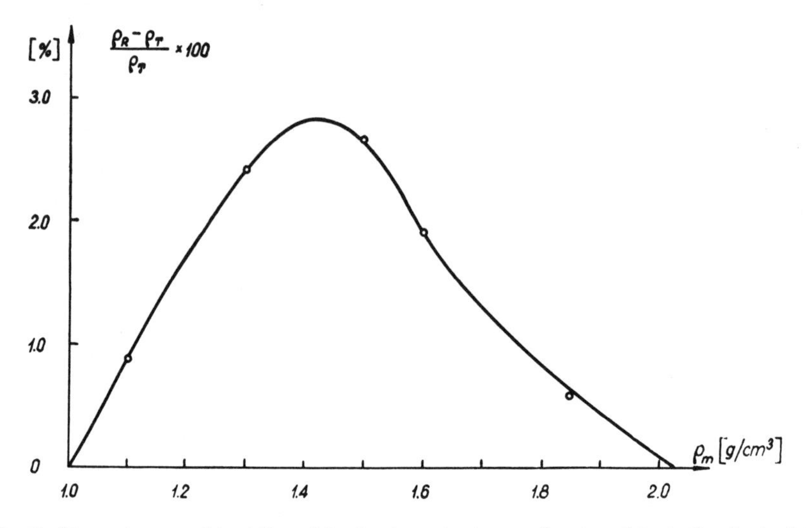

FIG.9. The kinematic error of the delivered density determination as a function of the in-line density [31].

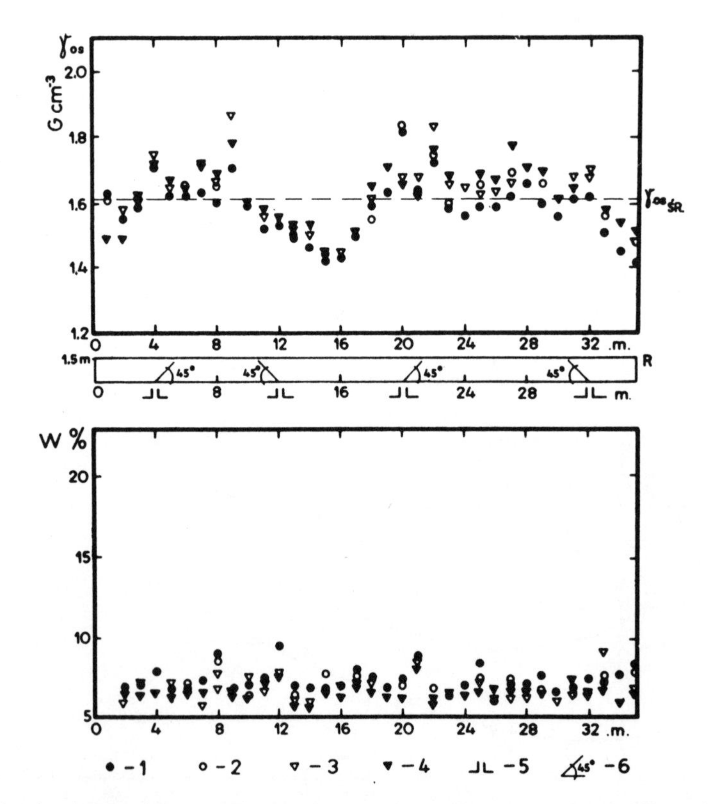

FIG.10. Local variations of density in the deposited back fill. Different points denote different days of measurements. R = filling-back pipe line, with an indication of the angle of the outlet inclination.

The same technique may be used to measure slip velocities. When the measuring base is large enough, or the multiscaler technique is applied, the slip velocity may be measured with high accuracy, without the need to label each component by means of different radioactive tracers. Figure 8 shows the results of measurements of the flow velocity of different granulometric fractions as a function of the mean density of the hydromixture [31].

Later, it was also possible to show that the kinematic error of the "cross-section" density measurement, e.g. by the radioisotope gauge, does not exceed 3% in the most unfavourable case (Fig.9).

III. DETERMINATION OF PHYSICAL PARAMETERS OF THE DEPOSITED BACK FILL

The mining back fill is transported hydraulically to the mining face and deposited there. Its role then is to minimize the vertical dislocation above the exploited mining area. It is desirable to have the compressibility of the back fill as low as possible. This, however, requires a better recognition of the technological parameters controlling the quality of the deposited back fill. Here, once again, radioisotope moisture and density gauges may become useful.

The problem of introducing a measuring gauge into the relocated back fill was solved by placing it into horizontal steel casing, 60 mm in diameter, a few tens of meters long, before the deposition of the back fill. Aluminium rods, 1 m long, were used to form a rigid tool to push the measuring probes into the casing [32].

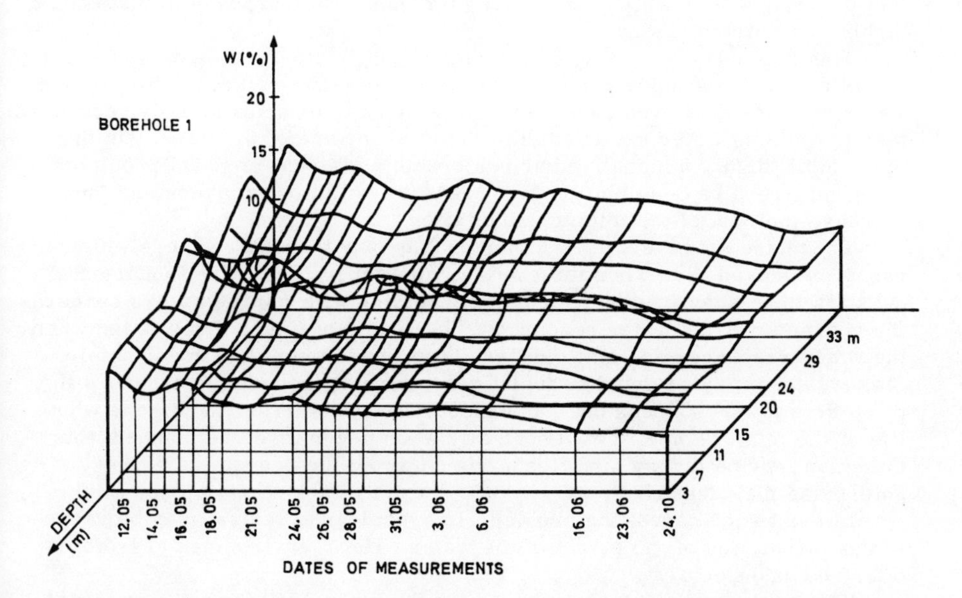

FIG.11. Variations of the moisture content with time in deposited back fill. Sample taken in borehole No. 1 from the Milowice coal mine, Sosnowiec, Poland in 1969.

Measurements performed by the Institute of Nuclear Techniques in Cracow, Poland, in one of the Upper Silesian coal mines, have shown that the average density of the deposited back fill is 1.6 g/cm^3. However, it was also demonstrated that, in the vicinity of the outlets of the pipe line, the kinetic energy of the outflowing hydromixture increases locally to a density up to 1.8 g/cm^3 (Fig.10). At the same time, the moisture content decreases exponentially from a value of approx. 15% volumetric H_2O to 5% H_2O after a few months (Fig.11).

Therefore, it was proposed that an appropriate modification of the technological process in the direction of the intentional use of the kinetic energy of flow may produce a significant reduction of the maximum compressibility of the back fill. From in-situ measurements of the density of the deposited back fill and a comparison with the value of the maximum obtainable density of the same material (obtained in the Proctor apparatus in the laboratory), one can determine the maximum expected compressibility. This knowledge could be of great importance in mining exploitation below inhabited areas.

It was also shown that the density of the hydromixture also influences the quality of the deposited back fill. It was demonstrated that the increase in density of a hydromixture to 1.5 g/cm^3, when compared with the one of 1.1 g/cm^3 commonly used, allows the potential compressibility of the back fill to be decreased from 12% to 7% [33].

IV. DETERMINATION OF ORIGIN OF MINING WATERS

The chemistry of waters in mines can be highly variable ranging from that of fresh waters which can be used for municipal purposes to that of highly concentrated brines.

The chemistry of mining waters significantly affects the costs of mining exploitation. If the mine pumps high-quality drinking water, or the pumped water contains dissolved rare elements which can be economically recovered, then the mining water contributes to the total income of the mine. On the other hand, highly mineralized brines create additional disposal problems. Often they require desalination to prevent the pollution of rivers and thus increase the cost of the mining exploitation.

There are, however, a few other problems with mining waters which cannot be solved even when the chemistry is understood. The fundamental question to be answered by hydrologists dealing with mining waters concerns the character of the water resources. The costs of future exploitation when the water resources are static differs from those when they are dynamic. Static waters may be dried out relatively easily by pumping, whereas with dynamic water resources one can expect continuous inflow during the whole life of the mine. Sometimes, especially when new mines are planned, this criterion may be a very significant one on which the economies of the enterprise may depend.

In most such cases, the problem, in hydrological terms, may be reduced to the recognition of the origin of the water. Here, environmental isotopes offer new prospects.

With young waters, interconnected with surface aquifers, the presence of tritium indicates recent replenishment of those aquifers. In many mines tritium measurements are performed. In Poland, e.g. in the Upper Silesian

Basin where coal deposits are exploited at depths between 100 and 1000 m, roughly half of the mines have tritium in their waters.

In salt mines a tritium indicator may be of special value. The presence of tritium indicates a connection with the source of contemporary water and this could warn against further exploitation in this direction because it shows that the long-term geohydrological equilibrium was disturbed and further exploitation may lead to sudden flooding of the whole mine.

To recognize the water relationships at greater depths, stable isotopes may be specially useful, namely deuterium and oxygen-18. Very old, so-called "relictic waters", are commonly contemporary with the age of the exploited deposit. They are mainly evaporated sea-waters. It is well known, on the basis of extensive experimental material [35], that sea-water is enriched in deuterium and oxygen-18 compared with meteoric waters (precipitation). On the basis of this regularity, oceanic water was chosen by hydrologists as a standard for determining the stable isotope composition

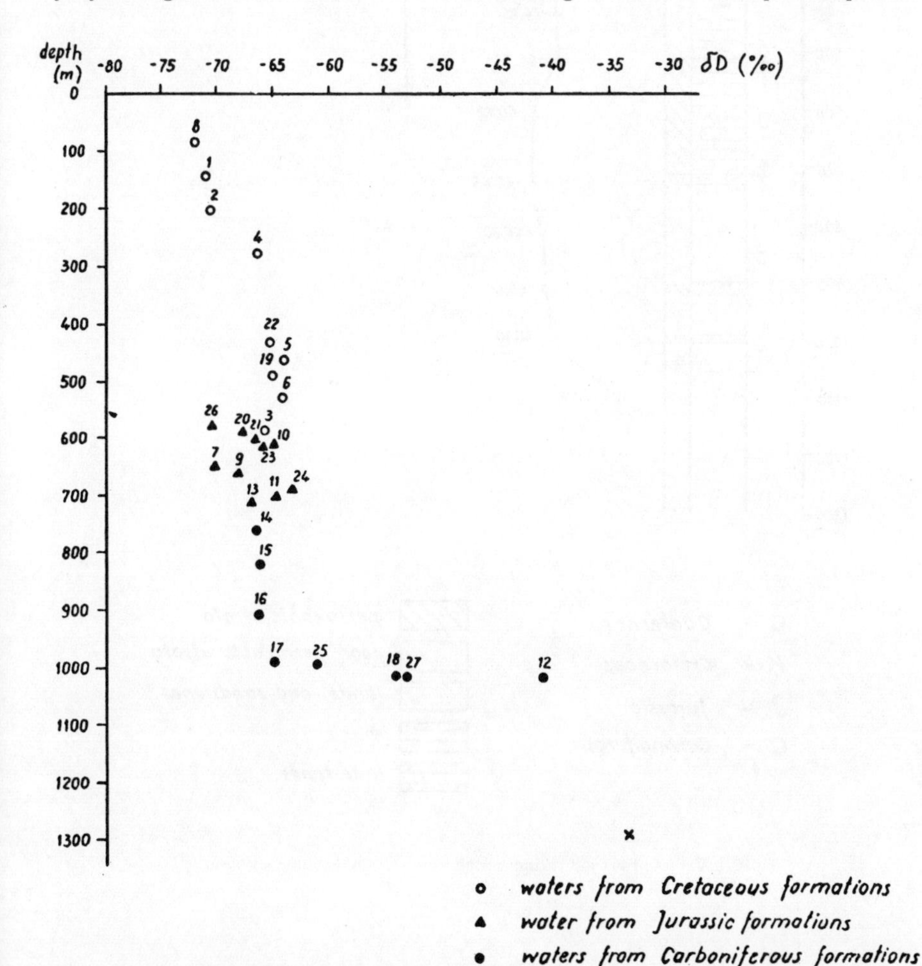

FIG.12. Value of δD plotted versus depth for Lublin coal basin waters.

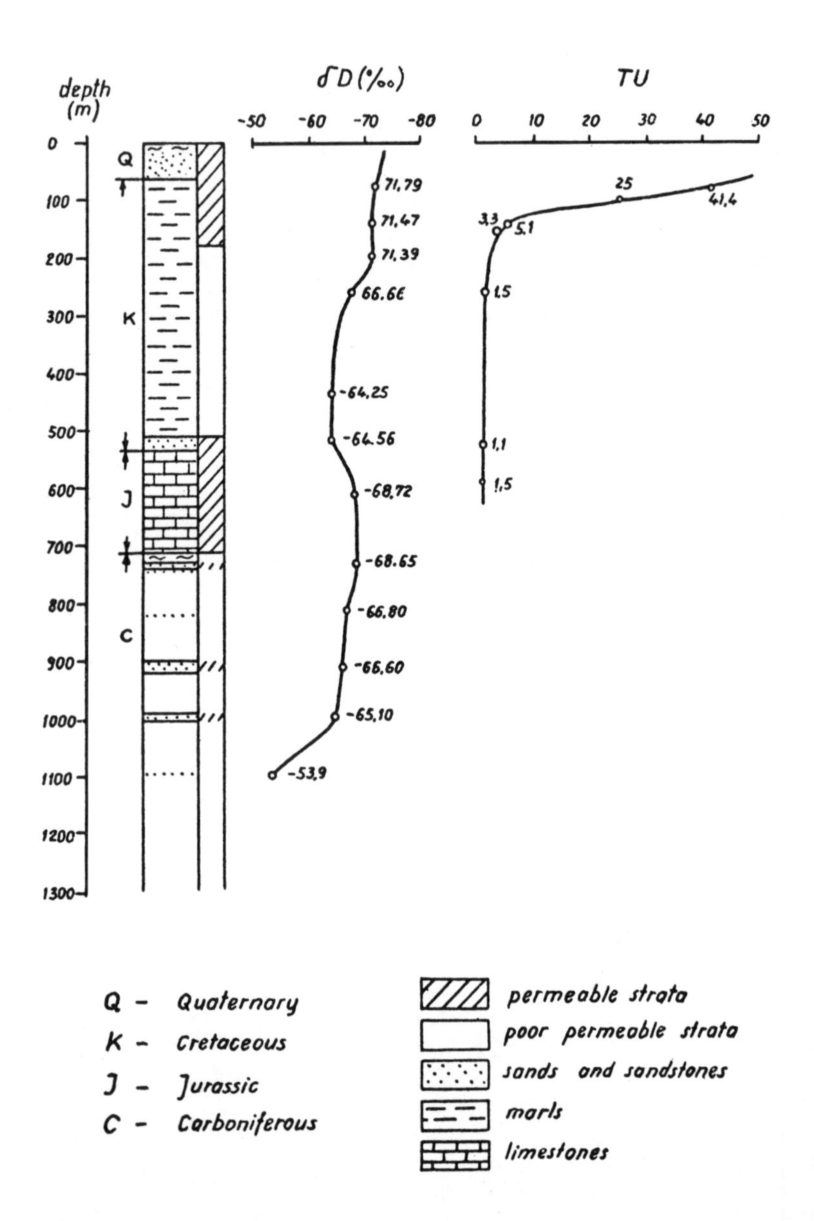

FIG.13. Values of δD and tritium in the waters of the Lublin coal basin.

(SMOW = Standard Mean Ocean Water). Groundwaters are usually depleted by approx. 2 - 10‰ SMOW composition in ^{18}O isotopes and 60 - 80‰ SMOW composition in ^{2}H.

There is a distinct difference in the stable isotope composition between relictic and meteoric waters. During millions of years of geological time, water of different isotopic composition migrated both upwards and downwards. As a result, one generally does not observe sudden changes in the stable isotope composition with depth, but rather a smooth transfer to more enriched values of D and ^{18}O up to a value close to SMOW.

There are, however, cases where this regularity is not preserved. When there are, or were in the past, separate interconnections and conditions reinforcing the flow, contemporary waters may penetrate quite deep and significantly affect this general picture. Such cases were recorded in the laboratory of the IAEA Isotope Hydrology Section in waters from the Lublin coal basin in Poland. Figure 12 shows values of δD plotted against depth. All the samples were collected from boreholes. It can be seen that the different meteoric waters ($\overline{\delta D}$ = 70‰ SMOW) reach a depth of 1000 m. Beginning from this depth, the water starts to be more enriched in deuterium (e.g. sample 12, depth 1060 m). The ^{18}O values show a similar picture for the same set of samples [34].

Figure 13 shows another set of data for the same site. Tritium-free water occurs at depths below 200 m. The deuterium values tend to decrease with depth. In the poorly permeable Cretaceous strata one can observe progressive enrichment in deuterium with depth until 520 m. Starting from this depth, permeable Cretaceous sandstones and Jurassic limestones begin which conduct contemporary waters from their outcrops (recharge area). As a result, at depths between 600 and 1000 m one can observe a zone of "freshing". For miners this means that at these depths they will deal with resources of "dynamic waters", predicting high continuous inflow to the mines.

Assuming enrichment in heavy isotopes with depth the same model is also proposed for use in oil prospecting. After seismic recognition of prospective oil-bearing structures, drilling normally follows. The first drillings do not always reach oil or gas, but always allow access to some deep waters. If these waters indicate a stable isotope composition enriched in heavy isotopes, this may indicate that the geological equilibrium in the whole structure is affected. In such cases, the chances to find oil or gas in the given structure are much higher than when meteoric waters penetrate into the investigated geological structure [36, 37].

V. FINAL REMARKS

The few examples of mining exploitation described in this paper indicate, in the opinion of the author, the applicability of methods based on nuclear techniques for investigating and recognizing the environment with which the miner is in contact every day.

REFERENCES

[1] SEGALIN, W.G., Application of radioactive isotopes to the automation of the coal industry (in Russian), Gostopthiekhizdat (1960).

[2] KASZKOWIAK, L., Milowice Coal Mine, Sosnowiec, Poland, private communication.

[3] Automatic colliery "Jan", Katowice, Poland, private communication.
[4] KORBEL, K., The applicability of radiogauges for measuring the density of flow solids-liquid mixtures, Isotopenpraxis 6 7 (1970).
[5] Nuclear Chicago Commercial Bulletin (1962).
[6] Colliery "Sosnowiec", Sosnowiec, Poland, private communication.
[7] Coal mine Milowice, Sosnowiec, Poland, private communication.
[8] BIERZIN, A.K., KUZNIEKOV, K.F., SULIN, V.V., BIELOV, U.I., VITOZHENC, G.C., MARTYNOV, I.T., SUSLOV, V.G., SHORNIKOV, S.I., "Gamma-activation analysis of elementary composition of rocks and ores", Radioisotope Instruments in Industry and Geophysics (Proc. Symp. Warsaw, 1965) 2, IAEA, Vienna (1966) 323.
[9] PRADZYNSKI, A., Copper determination in polymetallic ores, Warsaw 1969, unpublished.
[10] DARNLEY, A.G., LEAMY, C.C., "The analysis of tin and copper ores using a portable radioisotope X-ray fluorescence analyser", Radioisotope Instruments in Industry and Geophysics (Proc. Symp. Warsaw, 1965) 1, IAEA, Vienna (1966) 191.
[11] All Union Institute of Radiation Technique, Moscow, 1970, private communication.
[12] LEIPUNSKAJA, D.I., SAVOSIN, S.I., SULIN, V.V., Activation methods for elementary analysis of rocks and ores (in Russian), Viestnik Acad. Nauk, No. 1 (1972) 44.
[13] SAVOSIN, S.I., LEIPUNSKAJA, D.I., DYRKIN, V.I., ZEMTCHIKHIN, E.S., KRUTIKOV, M.G., "Practical aspects of the application of the neutron activation analysis in the USSR in prospecting and investigation of raw materials" (in Russian), Working paper on Practical Aspects of Neutron Activation Analysis, IAEA Panel, Vienna, Aug. 1973 (unpriced document).
[14] RHODES, J.R., DAGLISH, J.C., CLAYTON, C.G., "A coal-ash monitor with low dependence on ash composition", Radioisotope Instruments in Industry and Geophysics (Proc. Symp. Warsaw, 1965) 1, IAEA, Vienna (1966) 447.
[15] LJUNGGREN, K., "Review of the use of radioactive tracers for evaluating parameters pertaining to the flow of material in plant and natural systems", Radioisotope Tracers in Industry and Geophysics (Proc. Symp. Prague, 1966), IAEA, Vienna (1967) 303.
[16] MARTIN, T.C., HALL, J.D., MORGAN, I.L., "An on-line nuclear analysis system", Radioisotope Instruments in Industry and Geophysics (Proc. Symp. Warsaw, 1965) 1, IAEA, Vienna (1966) 411.
[17] KORBEL, K., "Some problems of the optimum design and exploitation of a hydraulic conveying installation for solid materials" (in Polish), Report of the Institute of Physics and Nuclear Techniques, Cracow, Poland, INT/55/1 (1974).
[18] WILSON, W.E., "Mechanics of flow with non-colloided inert solids", Proc. ASCE 67 (1946) 1434.
[19] KOTULSKI, V.V., Hydraulic theory of pressurized and unpressurized transport of a mixture of solid particles with water (in Russian), Gidrotekhnicheskoe Stroitelstvo, No 7 (1951) 417.
[20] WORSTER, R.C., Proc. Colloqu. on Hydraulic Transport of Coal, Nat. Coal Board, London, 1952.
[21] DURAND, R., CONDOLICS, E., "The hydraulic transport of coal and solid material in pipes", Proc. Colloqu. on Hydraulic Transportation, London, Nov. 1952.
[22] IVANOV, E.E., Displacement of grounds by means of pressurized and unpressurized streams, Izdatelstvo MRF (1952).
[23] NEWITT, D.M., et al., Hydraulic conveying of solids in horizontal pipes, Trans., Inst. Chem. Eng. 33 (1955) 93.
[24] VELIKANOV, M.A., Dynamics of bedded streams (in Russian), Gosud. Izd. Tekhn. Literatury, Moscow (1954).
[25] SMOLDYRIEV, A.E., Hydraulic and pneumatic transport at coal factories (in Russian), Myktiehizdat, Moscow (1956).
[26] PRZEWŁOCKI, K., Investigation of liquified solids by radiometric methods in colliery stowing pipelines, Scientific Bulletins, No. 119, Academy of Mining and Metallurgy, Cracow (1965).
[27] PRZEWŁOCKI, K., NIZEGORODCEW, P., Radiometric measurements of parameters of hydrotransport in industrial pipe lines, La Houille Blanche, No. 1 (1973) 59.
[28] MICHALIK, M., Density patterns of inhomogeneous liquids in industrial pipe lines measured by means of radioactive scanning, La Houille Blanche, No. 1 (1973) 53.
[29] KORBEL, K., Some problems of the optimization and control of press hydrotransportation (in Polish), Academy of Mining and Metallurgy, Cracow, Poland, No. 55/1/1974.
[30] KORBEL, K., MICHALIK, M., NIZEGORODCEW, P., PLECHTA, J., Measurements of parameters of hydrotransportation by means of radiometric methods, IAEA, Research Contract (1974).
[31] KORBEL, K., Investigation of flow parameters of a multifractional stowing mixture by means of radioactive isotopes, La Houille Blanche, No. 5 (1972) 427.

[32] BARANSKI, L., PRZEWŁOCKI, K., "Radiometric measurements of the bulk density and moisture content in a deposited back fill", Proc. Symp. Soil Mechanics and Foundation Engineering, Łódź, 1974.
[33] BARANSKI, L., PRZEWŁOCKI, K., Determination of the coefficient of the compressibility in deposited back fill by means of gamma-gamma and neutron-gamma measurements (in Polish), Przegląd Górniczy (in print).
[34] RÓŻKOWSKI, A., PRZEWŁOCKI, K., "Application of stable environmental isotopes in mine hydrogeology taking Polish coal basins as an example," Isotope Techniques in Groundwater Hydrology (Proc. Symp. Vienna, 1974) 1, IAEA, Vienna (1974) 481.
[35] Guidebook on Nuclear Techniques in Hydrology, Tech. Reports Series No 91, IAEA, Vienna (1968).
[36] ALEXEYEV, F.A., GOTTIKH, R.P., LEBIEDIEV, V.S., The use of nuclear methods in gas and oil geology (in Russian), Nedra (1973).
[37] Nuclear Geology (in Russian), Edition of the All Union Institute of Nuclear Geophysics and Geochemistry, Moscow (1974).

TRACERS

CONTROL OF MINERAL PROCESSING BY MEANS OF RADIOACTIVE TRACERS

K. LJUNGGREN
Isotope Techniques Laboratory,
Stockholm, Sweden

Abstract

CONTROL OF MINERAL PROCESSING BY MEANS OF RADIOACTIVE TRACERS.

Mineral processing involves a number of operations which, for technical and economical reasons, must perform efficiently. The use of radioactive tracers has contributed significantly both to the understanding of the physical and chemical phenomena which underlie the processes and to in-plant investigations of the transport dynamics of continuous flow processes. Applications to ore sorting and crushing, ore dressing, and transport in plant units are described.

1. INTRODUCTION

Processing of ore involves a number of operations which, for technical and economical reasons, must perform efficiently. For many of these operations, radioactive tracers have assisted in contributing towards a thorough understanding of the physical and chemical phenomena involved. This applies especially to flotation processes. In other instances they have been used in order to analyse and check the transport dynamics of continuous flow processes in actual plant operation. The technical and scientific literature bears witness to the widespread use of tracers in this field. In this brief review an attempt is made to distinguish between applications in different problem areas.

2. ORE SORTING AND CRUSHING

It has been suggested that ore could be separated from gangue by exploiting differences in the adsorption power for labelled substances. The rock would be passed through a solution of an active substance which is adsorbed either by the ore or the gangue; then the rock would be led past a measuring instrument coupled with a device for separating and collecting the ore [1]. There is no evidence that this application has been put into practical use.

Radioactive tracers have been used for determining grain size distributions [2] and for studying the change in grain size during grinding operations by adding labelled size-graded fractions of material [3 - 5]. Tracer techniques have further contributed experimental evidence for a theory of material breakage in milling operations. Several authors have used tracer techniques to determine the surface area of powdered materials.

Diamond imitations, made of an aluminium-cobalt alloy of exactly the same specific gravity as diamond, have been used to check the operation efficiency of gravity concentrators for diamonds [6].

The wear of grinding balls used for crushing ore in mills is important for the economy of the process. This wear has been studied by tagging a number of grinding balls either by adding radioactivity during manufacture or by irradiation. A number of radioactive balls were added to a mill, and recovered and weighed at regular intervals during the grinding operation [7].

3. ORE DRESSING

The surface sorption phenomena which underlie flotation processes have been studied by means of a variety of labelled substances by many investigators, among whom I.N. Plaksin and A.M. Gaudin and their co-workers occupy prominent positions. The reader is referred to available literature surveys for further information (see for instance Refs [8, 9]).

The distribution of a labelled substance that has been adsorbed onto a solid surface after it has been in contact with a solution of this substance can be determined with excellent resolution by a micro-autoradiographic technique in which the autoradiograph is often compared with a microphotograph of the surface in the same magnification in order to identify the various phases. This technique has been used for studies related to the flotation of sulphide ores such as galena, sphalerite, pyrite, pyrrholite etc. to obtain the distribution of collector agents (^{35}S-labelled xanthates) over mineral phases of different kinds. The influence of other variables such as pH, concentration of oxygen and xanthate chain length could be investigated, as well as the action of depressing agents (chromate ion) and enhancing agents (frothing agents, e.g. pine oil).

The adsorption of labelled anions and cations on quartz, fluorite, calcite and rare-earth minerals has been the subject of other investigations. Yet other studies refer to the flotation of metals, such as copper, silver and gold. The effect of cationic activation has also been investigated.

Quantitative studies of sorption on homogeneous surfaces can, with great advantage, be made by direct measurement of the amount of labelled substances on the surface. Owing to the high sensitivity of the tracer method it is possible to determine the thickness of the adsorbed layer as a function of, for instance, time and concentration in solution, even though the saturation thickness may be only a fraction of a continuous monolayer.

4. TRANSPORT DYNAMICS IN FLOW SYSTEMS

The conditions under which solids and fluids are transported through processing vessels of various kinds are of great technical and economical importance, and it is essential that the vessels possess the transport characteristics required by the process in question. These conditions of transport, more exactly the flow and macroscopic mixing processes, can be conveniently studied by observing the response of the system to a deliberately created concentration disturbance, a tracer signal [10]. Flow and mixing processes are generally independent of the concentration of the tracer and the traced substance and are therefore correctly reproduced by a tracer investigation. The usefulness of the tracer method for diagnosing continuous flow systems has in recent years been greatly increased by the development of new and efficient evaluation routines which also make the analysis of complex systems, e.g. systems with recirculation, a much simpler task.

Eichholz et al. [11] studied the transport of the material in a flotation circuit by using irradiated zinc ore and $^{64}CuSO_4$ as tracers (copper sulphate was used as a surface activator in the process). Various retention and contact times together with figures for the consumption of copper sulphate could be derived from the results.

Two investigations of the dynamics of flotation cells have been reported by Niemi [12, 13]. The mass transportation both in mechanically agitated and in pneumatic flotation cells was studied using $^{24}Na_2CO_3$ as a tracer for the fluid, irradiated tailings (producing ^{24}Na as the dominating activity at the time of measurement) for the gangue components and irradiated chalcopyrite for the floatable components. The experimental results were fitted to theoretically derived models and it could be shown that the residence time distribution of the pulp body in the mechanical cell corresponded to that of an ideal mixer in series with a small plug flow region. In this way the tracer investigations provided the essential transfer functions of the system.

Niemi [14] has also investigated flotation procedures of importance for the production of iron by using ^{59}Fe as the tracer nuclide. These investigations were followed by studies of the roasting of FeS matte.

Niemi discusses at some length in his papers the question of tracer representativeness which stands out as being particularly important in flotation studies when size-distributed solid material of complex composition has to be traced. Niemi has also attacked the problem of changing flow rate through a continuous flow system; this has led to a new type of representation for residence-time distributions [15].

King et al. [16] have described the dynamic testing of a mineral flotation cell in which a phosphate mineral was separated. Irradiated phosphate (^{32}P) and ^{137m}Ba (milked from ^{137}Cs) were used as tracers for the solid and aqueous phases, respectively. The authors stress the importance of having the extraction process operate continually at its maximum achievable level of efficiency in order to be able to recover ores from low-grade deposits in an economic way. This calls for a thorough understanding of the process. To improve information yield at low activity levels, digital filters for data smoothing were used. An alternative method of increasing information yield by smoothing experimental data has been suggested by Hansson et al. [17] who introduced Laguerre functions for representing residence-time distributions.

Dorr-type agitators in a pilot plant for leaching uranium ore were investigated for mixing efficiency by introducing tracers into the feed of the first tank in a series of four [18]. This investigation is one of the first to compare theoretically derived residence-time distributions for series-connected mixers with experimental data obtained with radioactive tracers.

The passage time for magnetite ore concentrate through a dephosphorizing plant has been measured using irradiated ore concentrate (^{59}Fe) as a tracer [19].

5. CONCLUDING REMARKS

This review of tracer applications for the control of mineral processing only presents those applications which are specific to this field. In addition, a number of standard tracer methods for flow-rate measurement, for leak detection, for analysis etc. are certainly useful and can easily be found in textbooks.

REFERENCES

[1] CHURCH, T.G., Can. J. Res. 28A (1955) 164.
[2] RAMDOHR, H., Kerntechnik 4 (1962) 318.
[3] HÜTTIG, G.F., SIMM, W., GLAWITSCH, G., Monatsh. Chem. 85 (1954) 1124.
[4] GAUDIN, A.M., SPEDDEN, H.R., KAUFMAN, D.F., Mining Engng 3 (1951) 969.
[5] FUSHIMI, H., Nucl. Engng 6 (1960) 19.
[6] NESBITT, A.C., "Radioactive diamond substitutes on South African diamond mines", Proc. Natl. Conf. Nucl. Energy, Appl. of Isotopes and Radiation, Pelindaba (1963) 105.
[7] KEYS, J.D., EICHHOLZ, G.G., "Measurement of the wear rate of cast grinding balls using radioactive tracers", Radioisotopes in the Physical Sciences and Industry (Proc. Conf. Copenhagen, 1960) 1, IAEA, Vienna (1962) 397.
[8] BRODA, E., SCHÖNFELD, T., The Technical Applications of Radioactivity 1, Ch. 7, Pergamon Press, Oxford (1966).
[9] INTERNATIONAL ATOMIC ENERGY AGENCY, Radioisotope Applications in Industry, Vienna (1963).
[10] LJUNGGREN, K., "Review of the use of radioactive tracers for evaluating parameters pertaining to the flow of material in plant and natural systems", Radioisotope Tracers in Industry and Geophysics (Proc. Symp. Prague, 1966), IAEA, Vienna (1967) 303.
[11] EICHHOLZ, G.G. et al., Trans Canad. Inst. Mining Met. 60 (1957) 63.
[12] NIEMI, A., A study of dynamic and control properties of industrial flotation processes, Acta Polytech. Scand. Chem. Metall. Ser. No. 48, Helsinki (1966).
[13] NIEMI, A., On the dynamics of a pneumatic flotation cell, Acta Polytech. Scand., Chem. Metall. Ser. No. 49, Helsinki (1966).
[14] NIEMI, A., "Tracer testing of particulate matter systems for their dynamics", Nuclear Techniques in the Basic Metal Industries (Proc. Symp. Helsinki, 1972), IAEA, Vienna (1973) 131.
[15] NIEMI, A., On theory and practice of tracer studies, Helsinki University of Technology, Control Engineering Laboratory, Otaniemi, Finland (1972).
[16] KING, R.P., WOODBURN, E.T., COLBORN, R.P., EDWARDS, R., SMITH, W.E., "Dynamic testing of numeral processing equipment using radioisotopes", Nuclear Techniques and Mineral Resources (Proc. Symp. Buenos Aires, 1968), IAEA, Vienna (1969) 117.
[17] HANSSON, L., KURTÉN, R., THÝN, J., The application of Laguerre functions for approximation and smoothing of count rates varying with time, to be published in Int. J. Appl. Radiat. Isot.
[18] TURGEON, J.C., Trans. Can. Inst. Min. Metall. 59 (1956) 14.
[19] Isotope Techniques Laboratory, unpublished report, Stockholm (1963).

ISOTOPE RATIOS

ECONOMICALLY IMPORTANT APPLICATIONS OF CARBON ISOTOPE DATA OF NATURAL GASES AND CRUDE OIL

A brief review

W. STAHL
Bundesanstalt für Geowissenschaften und Rohstoffe,
Hannover, Federal Republic of Germany

Abstract

ECONOMICALLY IMPORTANT APPLICATIONS OF CARBON ISOTOPE DATA OF NATURAL GASES AND CRUDE OIL: A BRIEF REVIEW.

Carbon isotope fractionations in hydrocarbons are briefly reviewed and examples of practical applications in the exploration of crude oil are given. Carbon isotope fractionations of natural gases are discussed. It is shown that the carbon isotope ratio of methane is predominantly determined by the environment (humic or sapropelic) and the maturity of its organic source material. In this way, isotope analyses of natural gases can be quantitatively used to characterize the maturity of their source rocks.

INTRODUCTION

In recent years carbon isotope analyses of insoluble (kerogen) and soluble (extract) organic carbon of probable source rocks, crude oil and gaseous hydrocarbons have been increasingly applied to economically important problems such as the correlation and identification of crude oil and source rocks. The aim of this paper is to give a short review on carbon isotope fractionation processes of hydrocarbons and their application in hydrocarbon research and exploration.

1. Correlation of crude oils

1.1. Isotopic maturity and age effect of solid and liquid hydrocarbons

The isotopic composition of organic matter is changed by maturity processes. Thermal stress changes the isotopic composition in favour of the heavier isotope because of the binding energy differences of isotopic molecules. This fractionation trend has been demonstrated in many laboratory experiments. For example, isotope fractionation during maturation of the immature oil schist of Messel (Bajor, 1969) is shown in Fig.1. In this figure increasing thermal stress is characterized by increasing values of C_R/C_T (Gransch and Eisma, 1966). The same tendency has been observed during thermal cracking of crude oil, Posedonian shale and Pennsylvanian coal by Sackett et al. (1972).

Figure 2 demonstrates the increase of ^{13}C in methane during the cracking process. The isotopic composition of methane changes in favour of the ^{13}C isotope with increasing amounts of methane, i.e. with increasing thermal stress. Moreover, the carbon isotope composition of the emerging methane

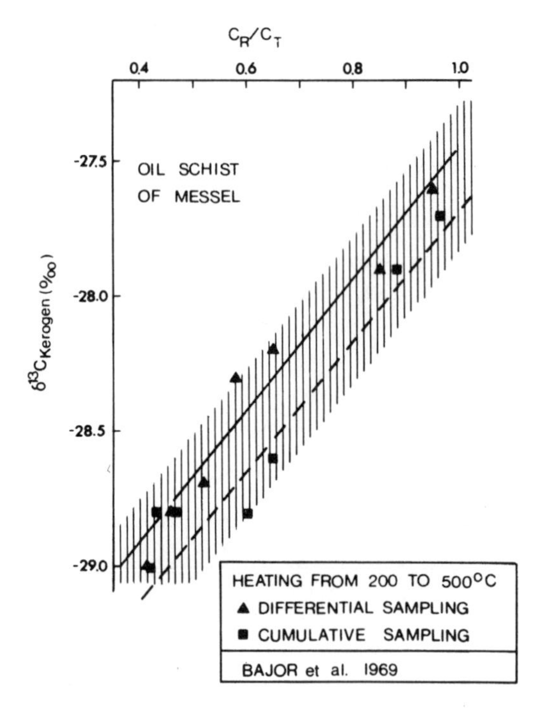

FIG.1. Carbon isotope fractionation of kerogen under artifical thermal stress.

$$\delta\,^{13}C\,(‰) = \frac{(^{13}C/^{12}C)^{\text{sample}} - (^{13}C/^{12}C)^{\text{standard}}}{(^{13}C/^{12}C)^{\text{standard}}} \cdot 1000$$

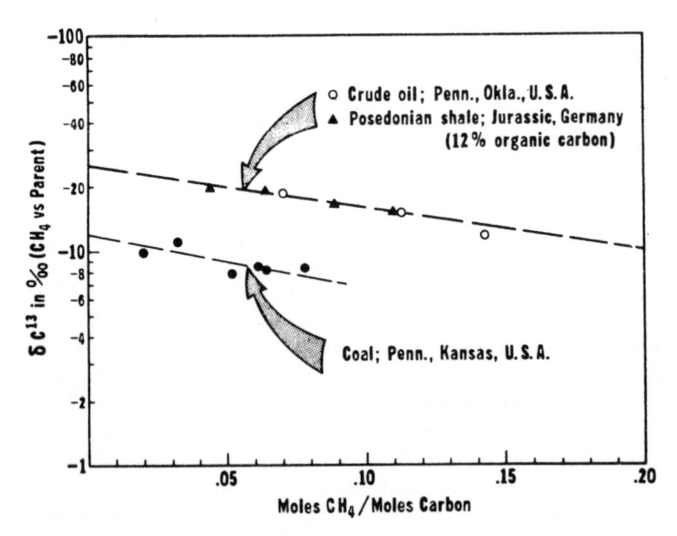

FIG.2. Carbon isotope composition of methane after artificial cracking of crude oil, Posedonian shale and Pennsylvanian coal (Sackett et al., 1972).

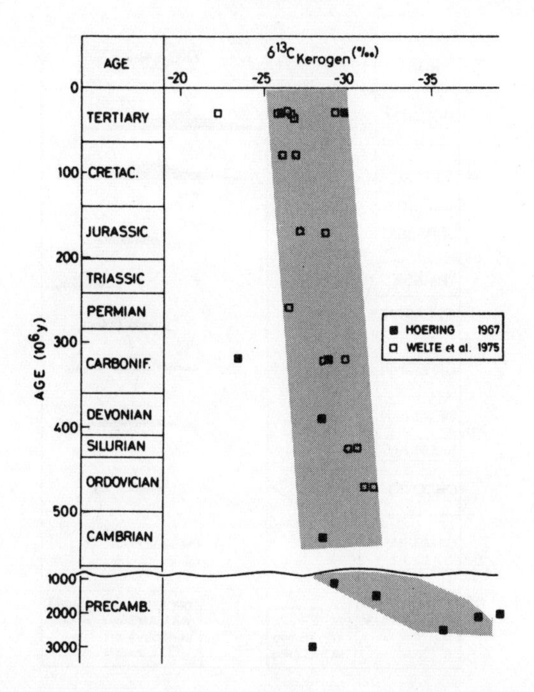

FIG.3. Change of the carbon isotope composition of kerogen samples during geologic time, with reference to PDB standard (after Hoerring, 1967; Welte et al., 1975).

is richer in light ^{12}C isotope than is its source material. In other words, the parent material must become enriched with heavy ^{13}C isotope during this artificial cracking process.

However, in natural samples, other fractionation processes, such as different intensities of photosynthesis, often superimpose themselves on this trend and reverse the effect. Measurements show that $^{13}C/^{12}C$ ratios of kerogen (Fig.3) and crude oil (Fig.4) generally decrease with geological time.

Welte (1970) discussed the relationship between the carbon isotope composition of crude oils and the amount of organic carbon in marine shales (Ronov, 1958) in different geological periods and showed that increasing intensity of photosynthesis could be responsible for the fractionation tendency observed in natural samples.

1.2. Crude oil correlation

Frequency distributions of carbon isotope ratios of crude oils in definite geological periods (Fig.4) are large probably due to differences in the isotopic composition of the organic material of their source rocks. In this way correlation of crude oils is possible, even if conventional methods, such as API gravity or the determination of the sulphur content, fail. As an example Fig.5 shows the distiction between different oil types in the Willistone basin, USA (Williams, 1974), by isotope data.

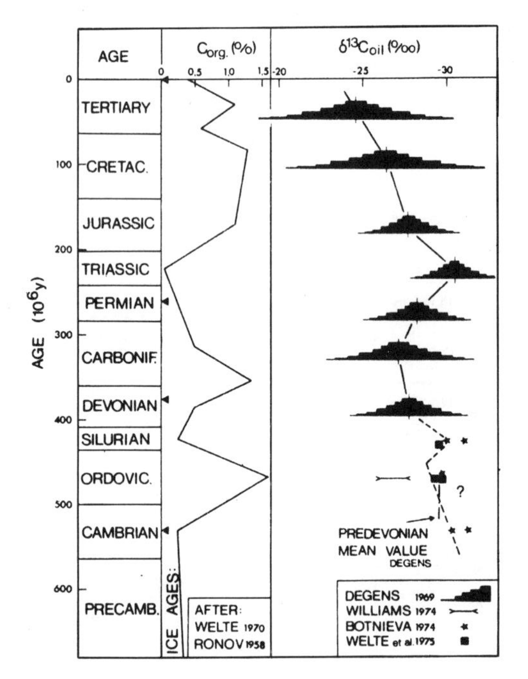

FIG.4. Change of carbon isotope composition (no PDB standard) of crude oils during geologic time (with reference to PDB standard).

It is obvious that the carbon isotope data of crude oils from the Willistone basin allow a clear distinction between type I and type II oils although other sets of data, like API gravity and sulphur content, overlap and cover the same range.

2. Source bed – crude oil correlation

Carbon isotope ratios of crude oils are equal, or slightly smaller, than those of the extract from a possible source rock. Figures 6 and 7 show source bed identification in the Willistone basin, USA (Williams, 1974).

Winnipeg shale was identified as the most probable source bed of the type I oils (Fig.6) and Bakken shale as the source rock of the type II oils on account of the very similar carbon isotope ratios of crude oils and the soluble organic matter of these shales. The Bakken shale, which could be a possible source rock of the type II oils, can be excluded by the carbon isotope composition of its extracts.

The isotope ratio of the extract (soluble organic matter) is smaller than that of the kerogen, i.e. the insoluble organic matter (Scalan, 1972). These relations are valid in Postdevonian times if oil, extract and kerogen are genetically linked and repeatedly allow for an identification of the source rocks or a correlation of crude oils and source rocks (Fig.8).

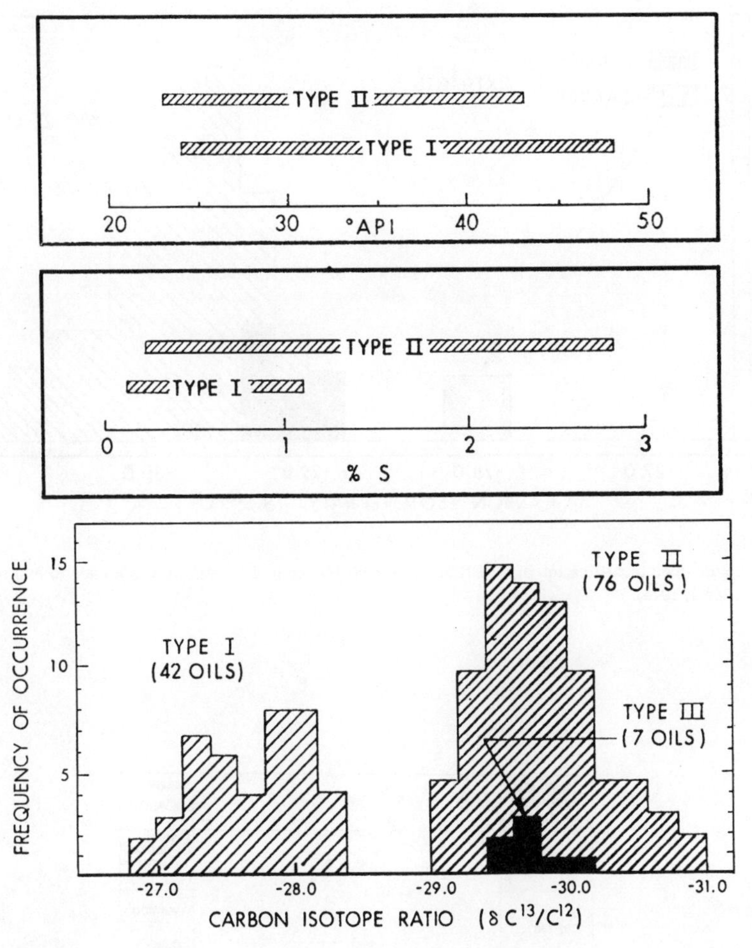

FIG.5. Comparison of API gravity, sulphur content and carbon isotope ratios of different types of crude oil in the Willistone basin, USA, with reference to PDB standard (Williams, 1974).

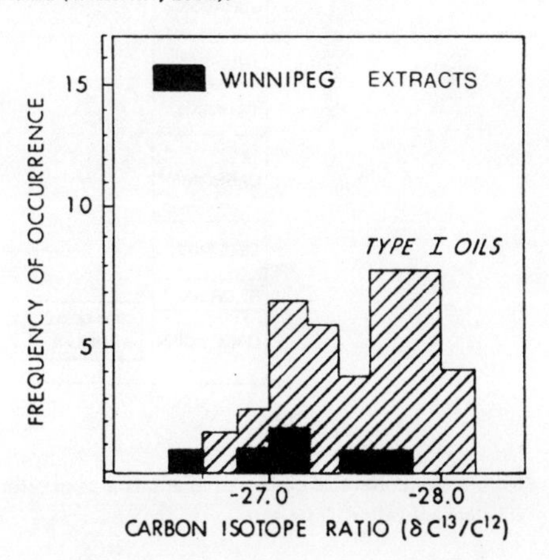

FIG.6. Source rock identification of type I oils in the Willistone basin, USA, with reference to PDB standard (after Williams, 1974).

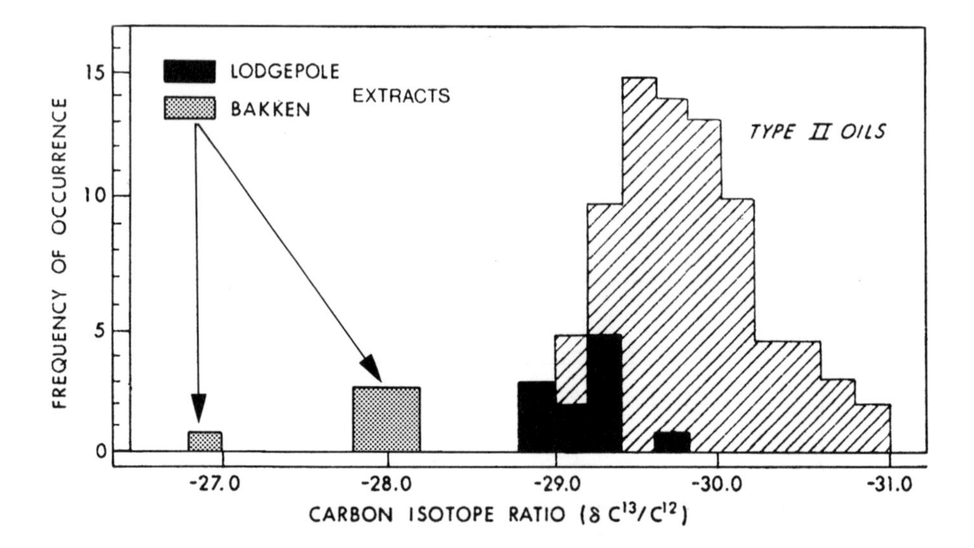

FIG.7. Source rock identification of type II oils in the Willistone basin, USA, with reference to PDB standard (after Williams, 1974).

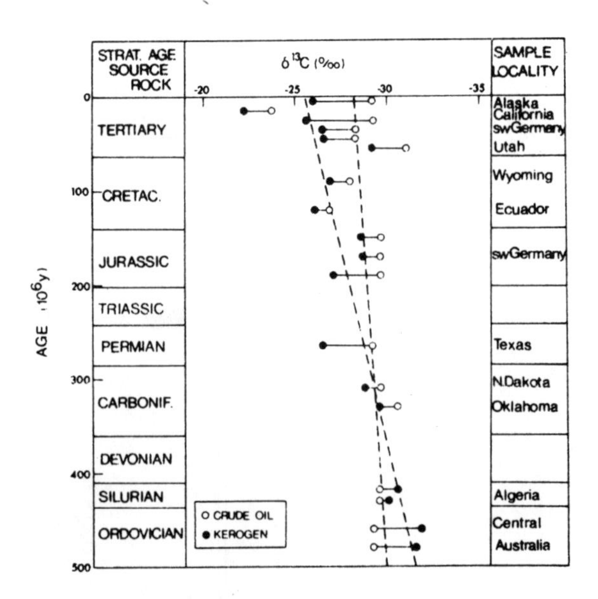

FIG.8. Age pattern of carbon isotope ratios of genetically linked kerogens and crude oils, with reference to PDB standard (after Welte et al., 1975).

3. Carbon isotope fractionations of natural gases

3.1. Isotopic equilibrium exchange fractionation between methane and carbon dioxide

Isotopic equilibrium (Craig, 1953) is only established in young natural gases which have originated from bacterial activity. Bacterial gases may thus be recognized by this effect.

3.2. Isotopic migration fractionation

At present there is no clear understanding of this phenomenon. Diffusion processes favour the light isotope; adsorption and solution effects seem to concentrate the heavier isotope in the gas. In some cases the data seem to indicate that the carbon isotope composition of methane in migrated natural gases was not significantly changed ($\lesssim 1‰$) by migration fractionation. Further investigations are urgently needed.

3.3. Nitrogen – δC_1 fractionation

Often a relationship between nitrogen content and carbon isotope ratios of methane can be observed (Colombo, personal communication, 1964, Stahl, 1970). High nitrogen contents are linked with isotopically heavy methane in natural gases. This fractionation is not yet fully understood but may clarify, in combination with the determination of nitrogen isotopes, the origin of nitrogen in natural gases (Stahl et al., in press).

3.4. Maturity fractionation

Molecules of various isotopes have different binding energies. Therefore, ^{12}C-^{12}C bonds are preferentially destroyed when organic material is exposed to thermal influence. Increasing temperatures will gradually destroy, in increasing numbers, ^{13}C-^{12}C bonds. Therefore, emerging gases, at the beginning of the maturity process of their source material, are isotopically light and become isotopically heavier with increasing maturity (cf. Fig.2).

4. Maturity lines

The relation of the hydrocarbon composition to the δC_1 values (C_1 = methane) of natural gases is called the "maturity line". This relationship allows for a distinction between methane derived from young source material and methane from thermocatalytically highly stressed organic matter (kerogen, oil, coal). There are different maturity lines for terrestrial and marine environments of source rocks. Although the relationship (cf. Fig.9) is only approximately valid in the high content range of higher hydrocarbons, it nonetheless allows for a rough estimation of the maturity of the source rocks (Stahl, 1975).

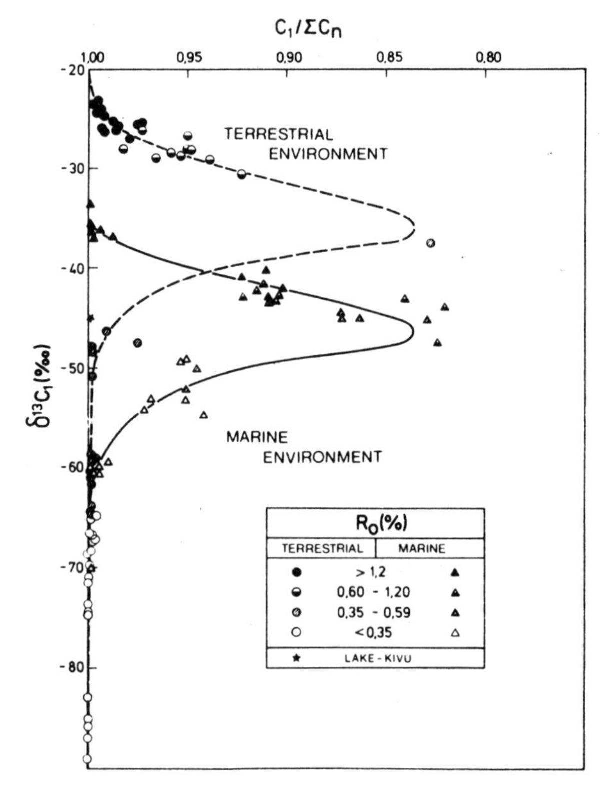

FIG.9. Isotopic maturity lines, with reference to PDB standard (Stahl, 1975).

5. Carbon isotope ratios of natural gases: maturity marks of their source rock materials

Carbon isotope ratios of methane (Stahl, 1974) and higher hydrocarbons (Stahl and Carey, in press) are related to the vitrine reflectance which characterizes the maturity of the organic source material. This relationship is demonstrated in Fig.10 by data from the USA and the Federal Republic of Germany.

Source rock maturities, determined by the carbon isotope composition of methane, can be verified by isotope analyses of gaseous higher hydrocarbons, like ethane and propane (Fig.11). In this way, mixing of bacterial and thermocatalytic gases in a reservoir rock or bacterial oxidation of the methane of a thermocatalytic gas can be recognized.

6. Carbon isotopes as a tool in hydrocarbon exploration

The isotope-maturity relationships (Figs 9, 10 and 11) are the basis for the application of carbon isotope exploratory work. It seems probable that in the near future we will be able to predict the maturity of source rocks and their approximate depths by isotope analyses of the migrating gases which can be sampled during exploratory drillings.

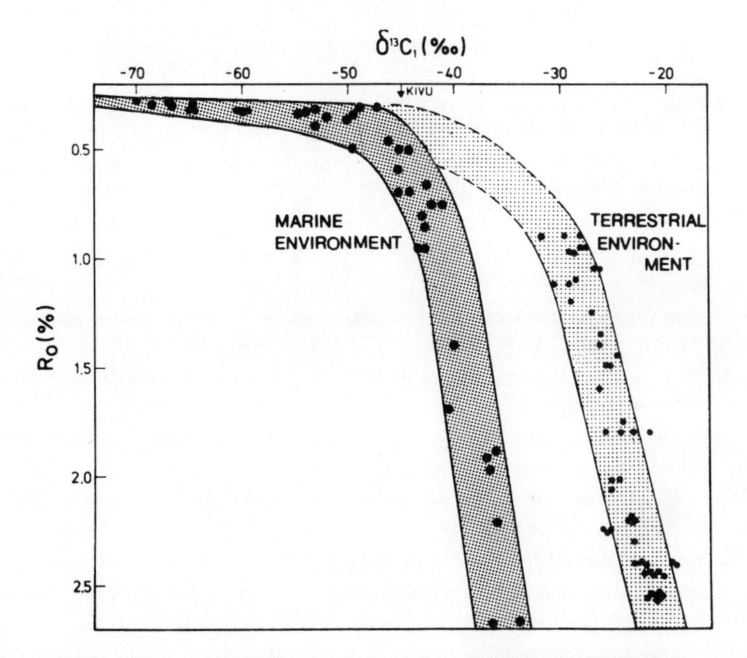

FIG.10. Carbon isotope ratios of methane of natural gases plotted against the vitrinite reflectances of the organic material of their source rocks, with reference to PDB standard (Stahl, 1975).

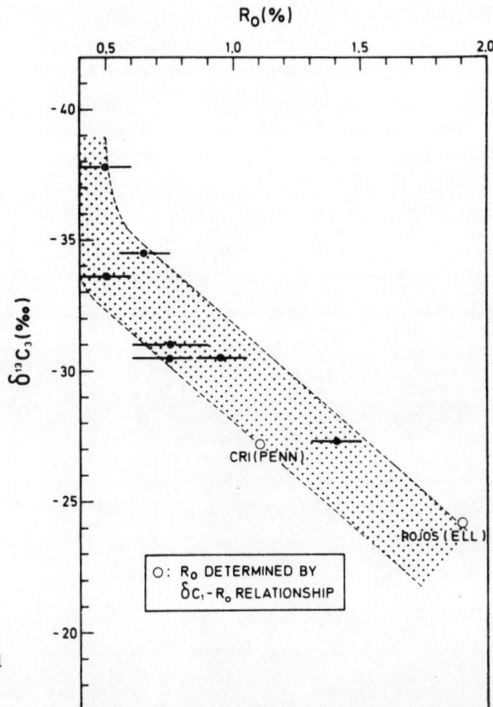

FIG.11. Carbon isotope ratios of propane of natural gases from west Texas plotted against the vitrinite reflectances of the organic material of their source rocks, with reference to PDB standard (Stahl and Carey, in press).

REFERENCES

BAJOR, M., ROQUEBERT, M.-H., VAN DER WEIDE, B.M., Transformation de la matière organique sédimentaire sous l'influence de la température, Bull. Centre Rech. Pau-SNPA 3 1 (1969) 113.

BOTNIEWA, T.A., GLOGOCZOWSKI, J.J., MÜLLER, E.P., Charakteristik unterschiedlicher Zusammensetzung der Kohlenstoff-Isotope in Erdölen verschiedenen Alters, Transl. from Nafta 30 No. 5 (1974) 199.

CRAIG, H., The geochemistry of stable carbon isotopes, Geochim. Cosmochim. Acta 3 (1953) 53.

DEGENS, E.T., "Biochemistry of stable carbon isotopes", Organic Geochemistry (EGLINTON, G., MURPHY, M.T.J., Eds), Springer-Verlag, Berlin (1969) 304.

GRANSCH, J.A., EISMA, E., Characterisation of the insoluble organic matter of sediments by pyrolysis: the C_R/C_T ratio, Paper presented at 3rd Int. Meeting Org. Geochem., London, 1966.

HOERING, T.C., "The organic geochemistry of Precambrian rocks", Researches in Geochemistry (ABELSON, P.H., Ed.), 2, John Wiley, New York (1967) 87.

RONOV, A.B., Organic carbon in sedimentary rocks (in relation to the presence of petroleum), Geochemistry 5 (1958) 510.

SACKETT, W.M., MENENDEZ, R., Study of the hydrocarbons and kerogen in the Aquitanian Basin, southwest France, Adv. Org. Geochem. (1972) 523.

SCALAN, R.S., Reservoir analysis, US Patent 3.649.201 — 14.3.1972.

STAHL, W., "Carbon isotope measurements on natural gases from north west Germany", Recent Developments in Mass Spectroscopy, Tokyo (1970) 718.

STAHL, W., $^{13}C/^{12}C$-Verhältnis norddeutscher Erdgase. Reifemerkmal ihrer Muttersubstanzen, Erdöl und Kohle 27 (1974) 623.

STAHL, W., Kohlenstoff-Isotopenverhältnisse von Erdgasen — Reifekennzeichen ihrer Muttersubstancen, Erdöl und Kohle 28 (1975) 188.

STAHL, W., CAREY, B.D., Jr., Source rock identification by isotope analyses of natural gases from fields in the Val Verde and Delaware Basins, west Texas, Chem. Geol. (in press).

STAHL, W., WOLLANKE, G., BOIGK, H., Carbon and nitrogen isotope data of upper Carboniferons and Rotliegend natural gases from north Germany and their relationship to the maturity of the organic source material, Paper presented at the 7th Int. Meeting on Org. Geochem., Madrid, 1975 (in press).

WELTE, D.H., Organischer Kohlenstoff und die Entwicklung der Photosynthese auf der Erde, Naturwissenschaften 57 1 (1970) 17.

WELTE, D.H., HAGEMANN, H.W., HOLLERBACH, A., LEYTHAEUSER, D., STAHL, W., Correlation between petroleum and source rock, Paper presented at World Petroleum Congress, Tokyo, PD3, Topic 5, May 1975.

WILLIAMS, J.A., Characteriztion of oil types in Willistone Basin, Am. Assoc. Petroleum Geologists Bull. 58 No. 7 (1974) 1243.

COMPARISON BETWEEN NUCLEAR AND NON-NUCLEAR TECHNIQUES

RELATION BETWEEN NUCLEAR TECHNIQUES AND OTHER PROSPECTING AND EXPLOITATION TECHNIQUES

J.F. CAMERON
Nuclear Enterprises Ltd.,
Reading, Berks,
United Kingdom

Abstract

RELATION BETWEEN NUCLEAR TECHNIQUES AND OTHER PROSPECTING AND EXPLOITATION TECHNIQUES.

In mineral prospecting and exploitation, a great variety of techniques is employed covering practically every physical and chemical method of qualitatively and quantitatively evaluating mineral-bearing materials. Nowadays, the most appropriate technique, or techniques, for a particular measurement is recognized and used, and often a combination of nuclear and non-nuclear techniques is employed. The main methods and their applications are summarized.

1. INTRODUCTION

In mineral prospecting and exploitation, a great variety of techniques is employed covering practically every physical and chemical method of qualitatively and quantitatively evaluating mineral-bearing materials. Nowadays, the most appropriate technique or techniques for a particular measurement is recognised and used, and in many instances, particularly in the exploration phase, a combination of nuclear and non-nuclear methods is employed.

Nuclear techniques generally have the advantages of providing a non-contacting quantitative measurement integrated over a significantly large volume of material, and the penetrating radiation often used enables measurements to be made through borehole liners or the walls of process vessels. In analysis, the nuclear methods generally measure physical properties and give little or no information on chemical properties. Potential hazards associated with the use of nuclear techniques are minimised by careful design and common-sense procedures, and seldom restrict their use.

The minerals to which nuclear techniques are mainly applied are oil and gas, coal and base metals. The oil and gas industries were among the first to use nuclear techniques routinely, and they are employed during exploration, completion of boreholes and the routine operation of wells. In the metals mining industry, they are used to locate and assess the value of ore bodies and in the subsequent mining and processing of ores.

2. PROSPECTING TECHNIQUES

Exploration for minerals can be considered conveniently in two stages:

(a) Regional reconnaissance

(b) Local detailed evaluation

2.1 Regional Reconnaissance

The purpose of the regional reconnaissance, or prospecting stage of mineral exploration, is to identify targets worthy of more detailed examination.

Economic oil and gas accumulations present relatively large targets compared with, for example, non-ferrous ore bodies, and once lithological and palaeontological examination of a sedimentary sequence has demonstrated the existence of potential source and trap rocks, airborne magnetic and surface seismic surveys are made to identify structural traps. These traps are then drilled systematically to determine the volume of the reservoir and the extractability of the hydrocarbon accumulation.

The problems of locating metal deposits are much more variable than those which arise in finding oil, gas and coal since circumstances which give rise to economically workable concentrations in the earth's crust are very rare. Most major metal deposits, excluding superficial placer and residual deposits, occur in geologically complex areas: that is, they have been formed at relatively deep levels in the crust in regions subjected to both tectonic disturbance and metamorphism, and it is not possible to define closely the situations in which mineral deposits are likely to occur. However, experience shows that there are some regions where deposits are more likely to be found, but the reasons are not clearly understood. Thus, exploration for metals is a much more difficult and speculative venture than is searching for hydrocarbons.

One of the principal aims of any regional reconnaissance program is to combine thoroughness with speed and low cost. A variety of remote sensing geophysical and geochemical methods is available at the present time[1,2] as summarised in Table I, but with a few obvious exceptions, such as the identification of large uranium or iron ore deposits, they are non-specific, and frequently produce a large number of anomalies that must be examined by other methods in order to ascertain whether or not they correlate with mineral deposits.

The term "remote sensing" as applied to mineral exploration is considered to cover all of the relevant detection and mapping methods which can be used from aircraft or spacecraft, and these are summarised in Table I. The established and most commonly used methods are aerial photography, airborne magnetic (AM), electromagnetic (AEM) and radiometric surveys, and these take a variety of forms which are under continual development. Recent developments include air sampling methods for regional geochemical surveys, colour infrared photography, very low

TABLE I. REMOTE-SENSING TECHNIQUES APPLICABLE TO MINERAL EXPLORATION

Magnetic (Aeromagnetic - AM)

(a) Total field (fluxgate, proton precession): ± 1 gamma

(b) High resolution total field (optical pumping): ± 0.1 gamma

(c) Vertical field (gyro, stabilized fluxgate): ± 25 gammas

(d) Vertical gradient (optical pumping)

Electromagnetic (AEM) - Non optical

(a) Inductive field continuous wave, audio frequency
Fixed wing rigid transmitter-receiver (T-R)
Fixed wing non-rigid T-R (bird)
Helicopter rigid T-R towed-bird
Helicopter rigid T-R aircraft mounted
Helicopter servo-alignment method

(b) Audio-frequency pulse system
Fixed wing induced pulse transient (INPUT)

(c) Radiated field
VLF radio-frequency
AF natural field

(d) Side-look airborne radar (SLAR): imaging

Optical

(a) Gas and vapour detection by spectral correlation

(b) Fluorescence by Fraunhofer line discrimination

(c) Far infrared spectral correlation (for rock identification)

(d) Photographic: imaging

Air sampling (Hg vapour, I_2, SO_2, aerosols etc.)

Radiometric

(a) Total gamma ray

(b) Gamma ray spectrometry

---

frequency (VLF) radio transmissions, optical-mechanical line-scanning infrared, side-looking airborne radar (SLAR), trace gas analysis by optical correlation methods and spectral scanning in the thermal infrared region.

Aeromagnetic surveys remain a prime tool in mineral exploration and, in addition to prospecting for magnetic ores, they help delineate geological features such as intrusive bodies

TABLE II. TECHNIQUES USED IN LOCAL DETAILED EVALUATION

Electromagnetic

Transmitter-receiver systems operating at VLF (horizontal and vertical loops)

Pulses EM

Induced polarisation (IP)

Self potential (SP)

Resistivity.

Magnetic

Fluxgate magnetometer

Proton precession magnetometer (1 gamma in fields of 20,000 to 100,000 gammas)

Cryogenic magnetometers (0.0002 gamma)

Gradiometers

Continuously recording ground magnetometer stations

Laboratory astatic magnetometers to measure magnetic remanence and susceptibility.

Seismic and Gravity

Geochemical

Atomic absorption and fluorescence

Optical emission and absorption

Zeeman modulated Hg spectrometer.

X-ray Fluorescence

Energy and wavelength dispersive techniques.

TABLE II (Cont.)

Borehole Logging
SP
IP
Resistivity
Focussed potential
Caliper
Inclination
Temperature
Mud flow rate and resistivity
Natural radioactivity (total and spectral)
Gamma-gamma (Compton and selective): density and analysis
Neutron-neutron: hydrogen, water and neutron absorbers (B, Cd etc.)
Neutron capture (prompt gammas): (Ni, Fe, Cu, Cr).

and major faults which often cannot be adequately covered by other methods. Most AM surveys are carried out with systems which measure the total magnetic field with an accuracy of ± 1 gamma. For these surveys the fluxgate magnetometer is being replaced by the proton precession magnetometer, and other magnetometers (Cs vapour and dual resonance - proton and electron spin) are under investigation. The oil industry is starting to use high resolution AM methods to determine subtle structural features in sedimentary basins where rock susceptibilities are low.

Low frequency inductive EM systems have achieved the most spectacular success in remote sensing of minerals and a number of rigid and non-rigid transmitter-receiver systems are carried by fixed wing aircraft and helicopters. These techniques can penetrate several hundred feet into the underlying terrain. To discriminate against conductive overburden, multiple component and induced pulse transient (INPUT) systems have been developed.

Radiometric surveys measure total gamma radiation and the intensity in certain energy regions characteristics of U, Th, and K. The latter provide reasonable estimates of the surface concentrations of these elements and provide complementary information in geological mapping, as well as delineating radioactive ore bodies. In some areas such as granite regions where rock susceptibility may be low, K counts can be high and structural detail appears on the radiometric maps which may be missing on AM maps. Haloes of hydrothermal alteration involving

TABLE III. TYPES OF LOGS FOR LITHOLOGICAL AND POROSITY INVESTIGATION IN THE OIL AND GAS INDUSTRIES

Mud and hole conditions	Data desired	Formation type		
		Unconsolidated formations (high porosity)	Medium formations (15 - 25% porosity)	Hard formations (low porosity)
Fresh muds (water base)	Lithology	Induction-electrical survey Electrical survey	Induction-electrical survey Electrical survey	Induction-electrical survey Laterolog Electrical survey Natural gamma ray
	Porosity	Sonic Gamma-gamma Microlog	Sonic Gamma-gamma Microlog	Sonic Neutron Microlaterolog
Salt muds (water base 20 000 ppm chlorides)	Lithology	Electrical survey Induction-electrical survey Laterolog Natural gamma ray, if necessary	Laterolog and natural gamma ray Electrical survey and natural gamma ray	Laterolog and natural gamma ray
	Porosity	Sonic Gamma-gamma Microlog	Sonic Gamma-gamma Microlaterolog	Sonic Neutron Gamma-gamma Microlaterolog
Oil-base muds	Lithology	Induction log and natural gamma ray		
	Porosity	Sonic, gamma-gamma and neutron		
Empty or gas-filled hole	Lithology	Induction log and natural gamma ray		
	Porosity	Gamma-gamma and neutron		
Cased hole	Lithology	Natural gamma ray		
	Porosity	Neutron		

potassium enrichment, for example, are characteristic of certain types of ore body, and radiometric anomalies can identify subsurface oil and gas reservoirs in certain geological environments.

Both analog and magnetic digital recordings are made of the sensor data as well as other relevant data such as position, altitude, attitude, speed and temperature, and computers are used to correlate and correct the data, and then plot contour or profile maps.

A new airborne geochemical technique is now being used to complement the geophysical techniques and help select targets for ground follow-up investigations. It depends on the phenomenon that metallo-organic compounds are dispersed into the overlying atmosphere by vegetation and soil humus. The optical analysis technique can detect Hg, Cu, Pb, Zn, Ni, Ag and also methane, carbon dioxide and sulphur compounds. Apart from correlation and screening, this method helps find deposits which are difficult to find by the other techniques, such as porphyry copper, lead, zinc and precious metals.

2.2 Local Detailed Evaluation

In following-up promising anomalies shown by regional surveys, an even greater variety of methods is employed, mainly on the ground, and these are summarised in Table II.

The main use of electromagnetic techniques is to determine the position and shape (dip angle, depth and dimensions) of conducting bodies to several hundred feet below the surface.

Magnetic methods are used for mineral and petroleum exploration, geological mapping, measurement of the magnetic properties of rocks or ferro-magnetic objects, paleomagnetism, conductivity mapping etc. Seismic and gravity assist in lithological studies.

Geochemical methods are applied mainly to analyse Hg and other base metals as an indication of base metal, precious metal and even uranium deposits, but also in general surveys for chemical anomalies. Energy and wavelength dispersive techniques are extensively used on the ground for single and multielement analysis of rock, soil and stream sediment samples and borehole cores.

Borehole logging is the main application of nuclear techniques in mineral surveys. This is a technique whereby instruments are lowered into a borehole so as to obtain direct information on the nature and composition of underground strata. Although perhaps the most accurate information can be obtained from a study of cores taken from a borehole, coring is both expensive and time consuming and good cores cannot always be obtained. The use of instruments which provide information from within the borehole, as well as providing data when cores are not available, also enables a check to be made on results from cores and makes it possible to use cheap, high-speed coreless drilling. The nuclear methods can be applied in both open and cased holes. Table III summarises the uses of different logs in oil and gas prospecting and exploitation.

3. NUCLEAR TECHNIQUES USED IN MINERAL EXPLOITATION

In the mining and processing of minerals, so many techniques are used that a brief survey of all methods is virtually impossible, so only a summary of nuclear instruments[3] is given below.

Oil and Gas Extraction

Borehole logging for interface detection, accurate perforation, location of cementation etc.

Coal Mining

Borehole logs to determine extent and thickness of seams and ash content.

Coal/Rock sensing (gamma-gamma) probes to control automatic coal cutting machines.

Level gauges for process control.

Density gauges to control sand-filling operations, heavy media separation plant etc.

Ash content gauges.

Belt weighers to measure and control mass flow in materials handling, blending etc.

Metal Mining

Borehole logs to control extraction and processing.

X-ray fluorescence analysers (mine working, laboratory and on-line) to control extraction and processing.

Level gauges for process control.

Density gauges, mainly on slurries to control processing.

Particle size analysers to control milling and grinding operations.

Belt weighers to measure and control mass flow to optimise extraction operations etc.

REFERENCES

[1] HOOD P., Mineral exploration: Trends and developments in 1973. Can. Min. J. Feb. 1974, pp. 163-214.

[2] BARRINGER A.R., Remote-sensing techniques for mineral discovery. Ninth Commonwealth Mining and Metallurgical Congress 1969. Mining & Petroleum Geology Section Paper 20. The Institute of Mining and Metallurgy, 44 Portland Place, London W1N 4BR.

[3] CAMERON J.F., CLAYTON C.G., Radioisotope Instruments. Pergamon Press, Oxford (1971).

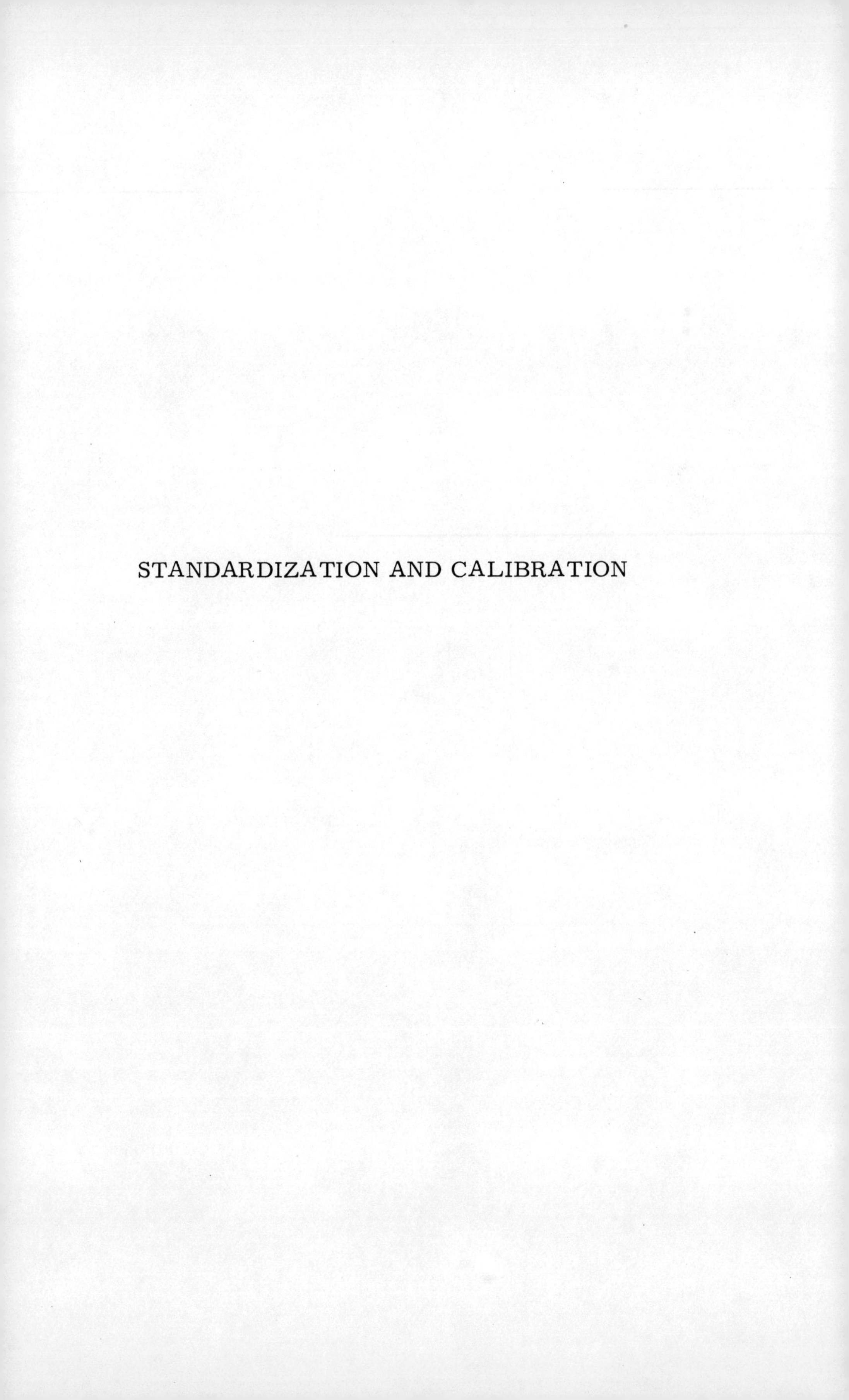

STANDARDIZATION AND CALIBRATION

EDXES FOR MULTIELEMENT TRACE ANALYSIS

Calibration standards and interlaboratory comparisons

J.R. RHODES
Columbia Scientific Industries,
Austin, Texas,
United States of America

Abstract

EDXES FOR MULTIELEMENT TRACE ANALYSIS: CALIBRATION STANDARDS AND INTERLABORATORY COMPARISONS.

A demand for calibration standards of a type not hitherto available has been caused by the rapid application of energy dispersive X-ray emission spectrometry (EDXES) to multielement trace analysis of solid particles, waters and biological materials. Also, interlaboratory comparison studies are necessary to test new instrumentation, new analytical procedures, and laboratories new to chemical analysis. This paper describes calibration standards recently made available, and two laboratory intercomparisons recently conducted in the USA. Further development of the standards is being done, and further intercomparisons are planned. Since multielement trace analysis of solid particles, waters and biological materials has importance in natural resource investigations, it was felt appropriate to report this work here.

1. INTRODUCTION

Applications of EDXES in the USA have grown rapidly over the past four years, particularly for multielement trace analysis of air particulates [1-14], waters [3,6,7,10,15] and biological materials [3,6-8,13,15]. Most applications to date are to pollution analysis and toxicology with some 20 groups active in the USA, conducting large-scale routine analysis programs, or research studies involving large analysis loads. We foresee a widespread application of the same technique in natural resource investigations such as geochemical, geobotanical and hydrogeological exploration, since the analytical requirements are similar. For example, in both pollution and geochemical analyses there is the same requirement to analyze very large numbers of samples for the same multielement combinations, and to perform the analyses rapidly, with high sensitivity and at low cost. The properties of the technique are equally appropriate; the determinations are non-destructive, rapid, and inherently multielement, with parts per million sensitivity in the analyzed specimen achievable for most elements of interest. The cost per analysis is very low and the number of samples routinely handled by one instrument and one operator is comparatively high (e.g., 10^3 to 10^4 samples/month for 10 to 40 elements/sample [4,12]).

Furthermore, the best results for all these analyses are obtained by presenting to the instrument a "thin specimen" usually in the form of a deposit on a membrane filter. The standards developed and under development, as well as the intercomparison studies completed and planned, all involve the "thin specimen technique" [1,2], so that all the information and technology developed and under development for pollution analysis is transferable to natural resource analysis. Adoption of the thin specimen technique in EDXES makes for maximum signal-to-background ratio; maximum elemental sensitivity in a given measurement time; a linear response of X-ray count rate to element mass per unit area over several orders of magnitude; minimization or complete absence of matrix absorption and enhancement effects; and reduction of particle size effects to a fairly straightforward calibration problem [10,16]. The thin specimen technique is also compatible with standard sample collection methods when the sample is particulate (e.g., aerosol, dust, soil, rock powder), and with methods of sample preconcentration when the sample is liquid (e.g., filtration, coprecipitation, ion exchange, evaporation, freeze drying).

2. CALIBRATION STANDARDS

Calibration standards are required in EDXES for various reasons such as: initial setting up of a multielement analysis job; periodic checking different kinds of drift; setting up correction factors for substrate self-absorption, deposit particle size, spectral overlap, etc.; and storing or updating library spectra for certain data analysis routines. Well-characterized samples are needed to periodically check the performance of individuals, techniques, instruments and laboratories. The well-characterized samples are of course standards whose composition is unknown to the party being tested.

2.1. Specification of Standards for Thin Specimen EDXES Analysis

In our experience the main requirements for an EDXES calibration standard are as follows:

1) The standard should have the same physical form as the analytical specimen, that is a deposit on, or impregnated in, a substrate such as a filter.

2) The material to be deposited should be easily and reproducibly prepared with an accuracy that is much better than the accuracy of the instrument to be tested. The composition should be known independently of any analytical method (i.e., made up from pure materials, gravimetrically or volumetrically) or it should have been independently analyzed many times by standard methods (i.e., a Standard Reference Material).

3) The mass per unit area of the deposit should be known at all points since different instruments look at different areas. Thus, the deposit should have a specified degree of homogeneity measured over specific dimensions.

4) The substrate should be "clean", that is contain negligible amounts of elements that would contribute to the spectral blank. It should be not too thick. The upper limit for particle-excited X-ray emission in this respect is about 4 mg/cm^2; that for light element analysis, whatever the exciting radiation, is about 1 mg/cm^2. It should be homogeneous, compatible physically and chemically with the deposit, and mechanically strong enough to be repeatedly handled.

5) Pure, single element standards are required for many of the more sophisticated data analysis routines, such as least squares fitting to library spectra. Multielement standards with non-interfering peaks are convenient for rapid setting up of multielement analysis jobs. Particulate standards, having a known particle size distribution

as well as element content, are required to calibrate particle size corrections particularly for elements below Fe in the Periodic Table. Multielement standards with preselected interferences are required for quality control procedures.

6) Not least, the reference standard must be stable with time and reasonably strong mechanically. It should not deteriorate at all during storage and should be essentially unaffected by the excitation X-rays or particles.

2.2. Description of Available Standards

The preferred basic standard is the "dried solution deposit". A gravimetrically and volumetrically prepared solution is deposited by the multidrop technique [17,18] on a wettable filter so as to uniformly impregnate it. The filter is then dried. The main advantages of using solutions are that they can be standardized very accurately gravimetrically or volumetrically, mixtures of many element combinations can be readily made, accurate dilutions are straightforward and the shelf-life of most stock solutions is very long.

Other methods have been tried including vacuum evaporation of metals or salts onto a substrate such as "Mylar", or gravimetric deposition of pure powders (e.g., Fe_2O_3) onto a filter. The main drawback of the vacuum evaporation method is that neither multielement nor particulate standards can be made. Also certain elements cannot be deposited. The required microgram quantities of pure powders, deposited on a possibly hygroscopic substrate, cannot be weighed accurately enough. However, particulate compounds containing fairly low percentages of the required elements are prepared this way since in this case milligram quantities are required and these can be weighed accurately.

The preferred substrate is a cellulose membrane filter such as Millipore type SMWP. Two diameters are used, 37 and 47 mm, together with two values of mass per unit area "regular" (about 4 mg/cm^2), for most medium and high Z elements, and "ultra thin" (about 1 mg/cm^2), for light elements. Cellulose fiber substrates (e.g., Whatman Type 541) are also used. They have a mass per unit area of about 9 mg/cm^2. Fiberglass and vinyl membrane filters are avoided because of high "blanks" of elements such as Zn, Ba and Cl. Plastic filters such as "Fluorpor" and "Nuclepore" are also unsuitable as substrates because they are not wettable by water. Other substrates such as purified quartz fiber also have been investigated.

Tables I, II, and III list single, two and multielement dried solution deposit standards that have at this time been prepared and tested [17]. The homogeneity, accuracy and reproducibility of these standards have been measured in two intercomparison programs [18] and in a number of laboratories in addition to the CSI laboratory. The absolute accuracy and reproducibility appear to be within the limits of error of the tests (about 5%). The homogeneity of deposit is also within these limits, over dimensions of a few mm. Shelf life of most standards is well over one year. Some exceptions exist (e.g., Sn, Br) where the substrate tends to disintegrate after a few months. Radiation damage appears to be minimal even from high power X-ray tubes. However, proton beams if too intense can burn small holes in the filters.

The dried solution deposits on cellulose filters satisfy the requirements for standards listed in Section 2.1 with the exception

TABLE I. SINGLE-ELEMENT DRIED SOLUTION DEPOSITS

Element	Nominal Conc. (μg/cm²)	Element	Nominal Conc. (μg/cm²)	Element	Nominal Conc. (μg/cm²)	Element	Nominal Conc. (μg/cm²)
Al(1)	100	Cr(2)	20	Se	20	Sb(2)	100
Si	100	Mn	20	Br(3)	20	Ba	100
P	100	Fe	20	Sr	20,100	La	100
S	75	Co	20	Zr(2)	10	Tl	20
Cl	50	Ni	20	Mo	20,40	Pb	20
K	50	Cu	20	Ag	100	Bi	20
Ca	40	Zn	20	Cd	100	Th	20
Ti(2)	10	Ga	20	In	50	U	20
V	20	As	20	Sn	100		

Notes: (1) Contains Hg (trace)

(2) Contains Cl

(3) Contains K in 1:1 atomic ratio to Br

TABLE II. TWO-ELEMENT DRIED SOLUTION DEPOSITS

Elements	Nominal Concentration ($\mu g/cm^2$)		Elements	Nominal Concentration ($\mu g/cm^2$)	
	On W541 or MSMWP	On M TAWP		On W541 or MSMWP	On M TAWP
Al(1)/Mn	200/50	50/15	Se/Mn	50/50	10/10
Si/Mn	400/50	100/10	Br/Mn(3)	50/50	15/15
P/Mn	300/50	70/15	Sr/Mn	50/50	10/10
S/Mn	150/50	40/15	Tl/Mn	50/50	15/15
Cl/Mn	120/50	30/15	Pb/Mn	50/50	10/10
K/Mn	80/50	20/10	Bi/Mn	50/50	15/15
Ca/Mn	80/50	20/15	Th/Mn	50/50	15/15
Ti/Mn(2)	50/50	15/15	U/Mn	50/50	10/10
V/Mn	50/50	15/15	Sr/Cd	100/80	25/20
Cr/Mn(2)	50/50	15/15	Zr/Cd(2)	10/80	2/20
Fe/Mn	50/50	15/15	Mo/Cd	40/80	10/20
Co/Mn	50/50	15/15	Ag/Cd	80/80	20/20
Ni/Mn	50/50	10/10	In/Cd	50/80	15/20
Cu/Mn	50/50	10/10	Sn/Cd	100/80	25/20
Zn/Mn	50/50	15/15	Sb/Cd(2)	100/80	25/20
Ga/Mn	50/50	15/15	Ba/Cd	100/80	20/20
As/Mn	50/50	10/10	La/Cd	90/80	20/20

Notes: (1) Contains Hg (trace)
(2) Contains Cl
(3) Contains K (1:1 atomic ratio w/Br)

TABLE III. MULTIELEMENT DRIED SOLUTION DEPOSITS

a) MLT-1

Element	Nominal Concentration (μg/cm^2)		Element	Nominal Concentration (μg/cm^2)	
	On W541 and MSMWP	On M TAWP		On W541 and MSMWP	On M TAWP
Al	50	15	Fe	15	4
K	20	5	Cu	2	0.5
V	3	1	As	2	0.5
Mn	8	2	Pb	6	1.5

Note: Hg (trace) and Br also present.

b) MLT-2

Element	Nominal Concentration (μg/cm^2)		Element	Nominal Concentration (μg/cm^2)	
	On W541 and MSMWP	On M TAWP		On W541 and MSMWP	On M TAWP
Al	30	7	Mn	5	1
S	10	3	Fe	20	5
K	30	7	Zn	2	0.5
V	5	1	Cd	40	10
Cr	8	2	Au	3	1

Note: Hg (trace) and Cl (excess) also present.

of particle size calibration. Particulate deposits are made by depositing the selected particulate material (after appropriate grinding and classification) on a preweighed filter using the "puff" technique [17], reweighing, then anchoring the particles by depositing a dilute solution of paraffin wax in benzene using the multidrop technique. This gives a known film of paraffin (about 40 $\mu g/cm^2$) whose X-ray absorption is small and can be corrected for. Drop tests are applied to check the adhesion of the deposit. Particulate standards of known mass per unit area and particle size are required mainly to apply correction factors in light element analysis (F to Fe). The most reliable standards are made using single component minerals and the following have been used: cryolite (for Na, F, Al), dolomite (for Mg, Ca), kyanite (for Al, Si), silica (for Si), pyrite (for Fe, S) and anatase (for Ti). Deposit weights from 0.3 to 10 mg, over 10 cm^2, are available, for all these minerals. Standard rocks such as G2, W2, BCR-1 and Sy-2 are multicomponent and may suffer segregation during deposition from air suspension. Canadian Syenite (Sy-2) was used in Intercomparison-II (see below) and some segregation of Fe, Mn and Ca is suspected at this time.

2.3. Current Development Work

Work is currently proceeding under contract with the U. S. Environmental Protection Agency. Refinements in deposition technique are under investigation, to obtain greater homogeneity of dried solution deposits and to avoid segregation of particulates deposited from air suspension. Deposition of particulates from water suspension, although ruled out earlier, is being reinvestigated.

Mixed particulate and dried solution deposits are being developed to provide multicomponent standards.

Preparation of thin plastic films from solutions or from melts, containing particulate or dissolved standard materials, is also being investigated.

3. INTERCOMPARISONS

Two intercomparison studies have been undertaken in the past 18 months both having the dual objectives of testing the EDXES technique against gravimetrically prepared "unknowns" and comparing EDXES with other techniques such as wavelength dispersive XRF, atomic absorption (AA), emission spectroscopy (ES) and neutron activation (NAA).

In the first intercomparison [18], 17 investigators each accepted a suite of 9 artificial, well-characterized samples. They consisted of either a 9-element dried solution deposit, or a standard rock powder, deposited on different substrates. The techniques used were EDXES (7 laboratories used X-ray excitation, 5 used particle excitation 4 of which reported data), wavelength dispersive (WD) XRF (2 laboratories, neither reported data), neutron activation (1 laboratory) and atomic absorption (2 laboratories, neither reported data), for a total of 12 groups reporting results.

The specific objectives were: to test the calibration and data analysis procedures of several new EDXES laboratories; to investigate the magnitude of particle size and other self-absorption effects in light element analysis; and to compare various substrates. The samples were monitored before and after analysis to check their integrity. The multielement dried solution deposits were satisfactory but the particulate deposits tended to fall off the substrate.

TABLE IV. INTERCOMPARISON-I, SELECTED RESULTS FOR EDXES

Number of Investigators Reporting:	11
Number of Investigators Included[a]:	8
Number of Elements Deposited:	9
Number of Elements Included in Data Evaluation:	8 (Bromine Proved Unstable)
Precision for K, V, Mn, Fe, Cu:	± 2.7% (On Both Filter Types)
Precision for Pb, Al, As (Whatman 41):	± 5.1%, ± 5.7%, ± 3.2% Respectively
(Millipore):	± 4.8%, ± 4.7%, ± 6.3%
Second Measurement by CSI After Return:	± 2% Avg.
Average Sample Uniformity for 8 Elements:	± 3.6%
Stability:	± 2.0%
Average Ratios:	
Solution Deposition/Gravimetric Standard for Whatman-41:	0.975 (Min. 0.90, Max. 1.07)
For Millipore:	1.026 (Min. 0.90, Max. 1.12)
Grand Mean Average Ratio for All Solution Samples:	0.998
7-Elements Average Standard Deviation:	3.8%

[a]Three Investigators were rejected whose data showed average calibration errors of -28%, -36% and + 40% for seven elements (excluding Al and Br).

Table IV summarizes the data obtained for the dried solution deposits by the 11 investigators who used EDXES. It is seen that, apart from 3 investigators who had systematic biases for all elements probably due to calibration errors, accuracies and precisions obtained were excellent. Measured areas of the standards varied from 0.5 to 5 cm^2 yet no bias was evident, indicating that the samples were homogeneous at least to within the measurement accuracy of the investigators.

The second Intercomparison included 35 laboratories (31 USA, 4 non-USA) utilizing the following techniques: X-ray excited EDXES (11 groups), particle-excited EDXES (7), WDXRF (8), atomic absorption (7), neutron activation (5) and emission spectroscopy (3).

Each laboratory received a pre-monitored suite of eight samples, consisting of pairs of the following four types. Type A was a 10-element dried solution deposit on a Millipore SMWP substrate containing the following elements in standardized amounts in the range 1 to 100 $\mu g/cm^2$, Al, S, K, V, Cr, Mn, Fe, Zn, Cd and Au, together with excess Cl and traces of Hg.

Type B was a weighed particulate deposit on the same substrate, consisting of about 300 $\mu g/cm^2$ of an SRM rock containing known concentrations of Na, Mg, Al, Si, K, Ca, Ti, Mn, Fe and traces of other elements. The average particle size was approximately 20 microns and the size distribution was available. Type C was a real air particulate sample collected by U. S. Environmental Protection Agency personnel from a fossil fuel power plant. The fourth group consisted of appropriate blanks.

The specific objectives of Intercomparison-II are as follows: Firstly, to evaluate various methods of correcting EDXES for its main sources of error, namely line-line and line-background "spectral" interferences, self absorption of the substrate and deposit for light element X-rays, and particle size effects for light elements. Secondly, to compare EDXES with the more established techniques namely WDXRF, AA, NAA and ES, for analysis of deposits. Thirdly, to test the usefulness and accuracy of the CSI standards which comprised the well-characterized unknowns.

At this time, which is just after the final deadline for data, some 22 investigators have turned in results and preliminary statistical analyses have been made. Publication is intended in "X-Ray Spectrometry" in 1975.

The following observations are worthwhile:

1. The EDXES data is reasonably accurate with a high percentage of all the possible determinations actually reported. Also the data was turned in promptly. The non-EDXES data is sparse and was turned in relatively late. Only a few of the possible determinations were reported and then relatively inaccurately. For example, in the case of the dried solution deposits 128 out of 160 possible results (16 investigators, 10 elements) were reported by EDXES groups with + 0.2% bias and 11% overall relative standard deviation. For all the other techniques 24 out of 50 values were reported, with -8% bias and 22% over-all relative standard deviation. In the case of the rock powders, for the main elements Al, Si, K, Ca, Ti, Mn and Fe, the 14 EDXES investigators reported 82 out of a possible 98 determinations. The overall bias was + 5% and the relative standard deviation 11%. The 5 non-EDXES investigators reported 14 out of 35 possible determinations; their overall bias being -27% and relative standard deviation 61%.

2. The sample sets produced for the intercomparison appeared to satisfy the requirements discussed in Section 2.1 with the exception that some segregation of Ca, Mn and Fe might have occurred during deposition of the rock powder from air suspension.

3. The various methods used for unfolding unresolved peaks in EDXES seemed to work satisfactorily. This is significant since the spectral interferences in the dried solution deposits were deliberately severe and included SK + ClK + AuM, KK + CaK + CdL, ZnK + AuL, VK_β + CrK_α, CrK_β + MnK_α and MnK_β + FeK_α.

4. Correction factors for light element X-ray absorption in the substrate, and for particle size and self-absorption effects in the deposits were made in some cases, and the improvement in accuracy was evident. However, work still needs to be done in this area.

Finally, it has been agreed at least among those who took part in the above work that intercomparisons of this nature should continue so that laboratories new to the EDXES technique can test their methods and so that new analytical procedures can be evaluated.

REFERENCES

[1] RHODES, J. R., PRADZYNSKI, A. P., SIEBERG, R. D., FURUTA, T., "Applications of Low Energy X- and Gamma Rays" (C. A. Zeigler, Ed.), Gordon and Breach (1971) 317.

[2] RHODES, J. R., ASTM Special Technical Publication 485, (J. C. Russ, Ed.) (1971) 243.

[3] JAKLEVIC, J. M., GOULDING, F. S., IEEE Trans. Nucl. Sci. 3 (1972) 384.

[4] CAHILL, T. A., UCD-CNL 162, University of California (1972).

[5] RHODES, J. R., PRADZYNSKI, A. H., HUNTER, C. B., PAYNE, J. S., LINDGREN, J. L., Environmental Science and Technology 6 (1972) 922.

[6] JOHANSSON, T. B., AKSELSSON, R., JOHANSSON, S. A. E., "Advances in X-Ray Analysis Vol. 15", (K. F. J. Heinrich, C. S. Barrett, J. B. Newkirk and C. O. Ruud, Eds.) Plenum Press (1972) 343.

[7] GIAUQUE, R. D., GOULDING, F. S., JAKLEVIC, J. M., PEHL, R. M., Anal. Chem. 45 (1973) 671.

[8] COOPER, J. A., Nucl. Instrum. Meth. 106 (1973) 525.

[9] BONNER, N. A., BAZAN, F., CAMP, D. C., UCRL-51388, University of California (1973).

[10] RHODES, J. R., Amer. Lab. 5 (July 1973) 57.

[11] GILFRICH, J. V., BURKHALTER, P. G., BIRKS, L. S., Anal. Chem. 45 (1973) 2002.

[12] PAYNE, J. S., "Analysis Instrumentation Vol. 11", (A. H. Keyser, A. M. Bartz and F. Combs, Eds.) Instrument Soc. Amer. (1973) 152.

[13] ONG, P. S., CHENG, E. L., SROKA, S., "Advances in X-Ray Analysis Vol. 17" (L. Grant, C. S. Barnett, J. B. Newkirk and C. O. Ruud, Eds.) Plenum Press (1974) 269.

[14] MORRISON, J. F., ELDRED, R. A., Ibid. 560.

[15] WALTER, R. L., WILLIS, R. D., GUTKNECHT, W. F., JOYCE, J. M., Anal. Chem. 46 (1974) 843.

[16] RHODES, J. R., HUNTER, C. B., X-Ray Spectrometry 1 (1972) 113.

[17] PRADZYNSKI, A. H., RHODES, J. R., 25th Pittsburgh Conf. on Analytical Chemistry and Applied Spectroscopy, Cleveland, March 1974.

[18] CAMP, D. C., COOPER, J. A., RHODES, J. R., X-Ray Spectrometry 3 (1974) 47.

STANDARDIZATION IN NEUTRON ACTIVATION ANALYSIS OF GEOCHEMICAL MATERIAL

H. PUCHELT, U. KRAMAR
Institut für Petrographie and Geochemie
der Universität Karlsruhe,
Karlsruhe,
Federal Republic of Germany

Abstract

STANDARDIZATION IN NEUTRON ACTIVATION ANALYSIS OF GEOCHEMICAL MATERIAL.

Different methods of standardization in neutron activation analysis are described, namely by means of a single-flux monitor, a multi-comparator, single-element standards and geochemical reference materials. The applicability of these methods for geochemical analysis is discussed. The possibility of errors due to flux variations is shown for the flux-monitor method. The source and availability of five new geochemical reference materials (granite, latiandesite, andesite, phonolith and hornblende) are given.

INTRODUCTION

The analysis of geochemical samples generally involves a large number of elements. In many laboratories these investigations, which are based on activation with thermal or epithermal reactor neutrons, are performed by instrumental analysis using multichannel analysers and Ge(Li) detectors. The complex composition of the matrix with varying concentrations of isotopes with different half-lives requires special precautions in the preparation of standards. The following ways of calibration are possible:

1. By using a single-flux monitor as an indicator of the neutron flux to which the sample was exposed. Cobalt wires, gold foils and gold-seized aluminium wires are mainly used for this purpose.
2. By applying multi-flux monitors, e.g. the triple comparator method of Van der Linden et al. (1974) by which the flux parameters for the thermal and epithermal range are obtained from three different suitable monitors (^{97}Ru, ^{103}Ru, ^{105}Ru).
3. By means of single-element standards which contain one or a few of the elements to be determined in the unknown materials in comparable amounts.
4. By means of geochemical reference samples, e.g. the reference rocks of the United States Geological Survey and many others (Flanagan, 1974).

The main requirement of each standard is that it allows the exact determination of each element desired. The following conditions have to be satisfied:

The neutron flux through the sample and reference have to be the same and must be known. The neutron flux must not be changed by the introduction of either the sample (including its container) or the standard.
Self-shielding must not occur in either the standard or the sample.

METHODS OF STANDARDIZATION

The use of flux monitors allows the determination of the integral neutron flux at a particular observation point within the reactor, but the application of a single-flux monitor is based on several strongly over-simplified assumptions. For example, the ratio of thermal flux to epithermal flux is assumed to be constant since the epithermal part is considered only when it is included in the cross-sections that are valid for reactor neutrons (see the Tables for neutron cross-sections in Lederer et al. (1967) and Seelmann-Eggebert et al. (1974)). Because the ratio of epithermal to thermal neutrons is different for each reactor, the neutron cross-sections must be determined experimentally for each particular condition of irradiation. All simplifications made for the triple-flux monitor are also valid for the single-flux monitor.

Satisfactory results with the single-flux monitor method can be expected for isotopes with cross-sections that almost follow the 1/v law.

Triple-flux monitors

With triple-flux monitors the integral epithermal and thermal fluxes can be obtained from the thermal cross-sections and the resonance integrals (Van der Linden et al., 1974). The values for the resonance integrals reported in the literature differ considerably, e.g.: for ^{168}Yb, I_0 = 23040 b (Van der Linden et al., 1974), 14700 b (Steinnes, 1971), 14700 - 35700 b (a total of

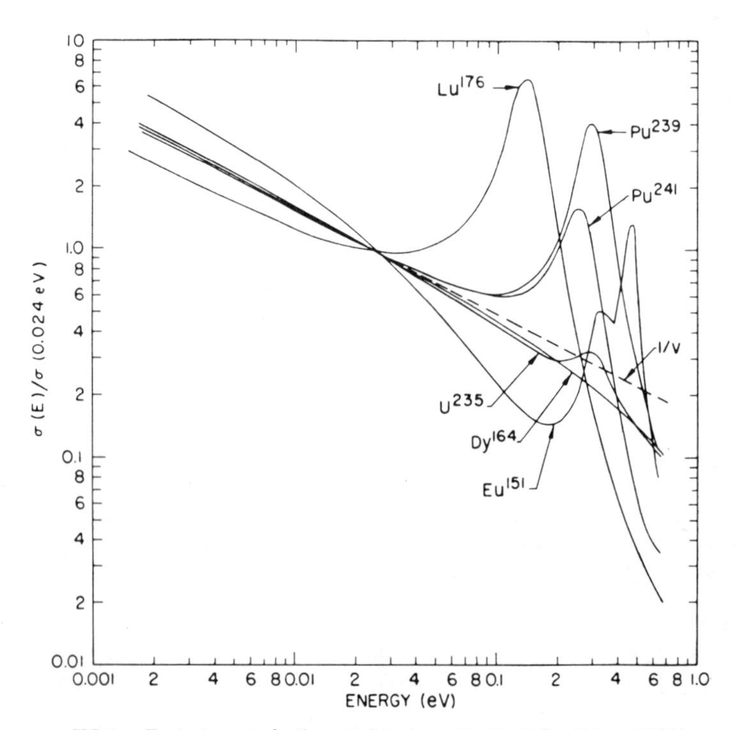

FIG.1. Typical curves for thermal detector activation (after Volpe, 1968).

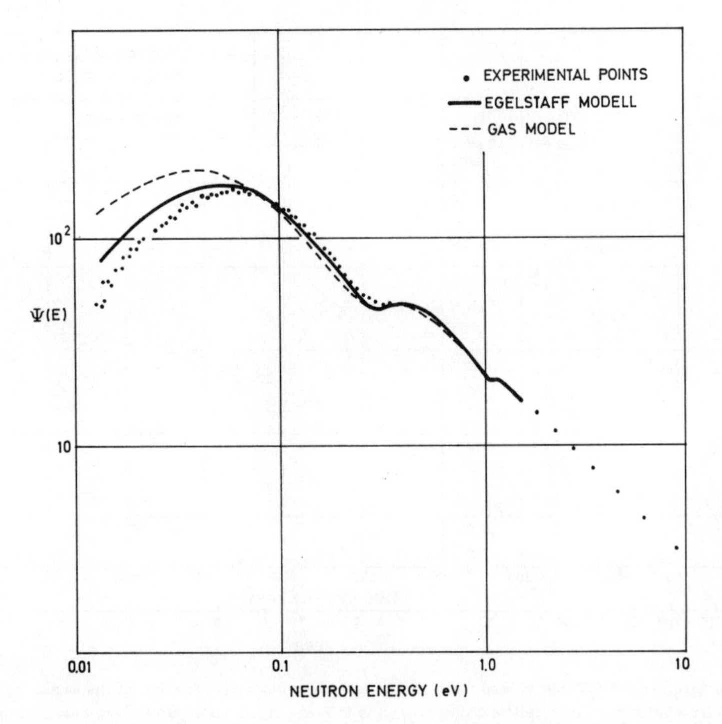

FIG.2. Theoretical and experimental spectra in core 7 of the Zenith reactor (after Beyster and Neill, 1968).

6 values from IAEA TRS No. 156, 1974), and for ^{151}Eu (product ^{152g}Eu) the values are: 5564 b (Van der Linden et al., 1974), 3847 b (Steinnes, 1971), 3550 and 11410 b (IAEA TRS No. 156, 1974). This list of examples can easily be expanded from the references in the papers cited.

These differences in the values for resonance integrals may be due to resonances in the transition range between thermal and epithermal neutron velocities. As an example, Fig. 1 gives the characteristics for ^{176}Lu and ^{151}Eu which show resonances well below 1 eV.

Even in the triple-flux monitor method several simplifications are made: it is assumed that the spectrum of the thermal neutron is shaped to fit a Maxwellian distribution with a maximum at 2 to 3°C above the moderator temperature (Day, 1968). For the epithermal part a continuous distribution is assumed which approximately corresponds to the 1/v law. In the transition range between thermal and epithermal the actual neutron spectrum shows the depression of the flux at 0.3 eV and 1 eV (Beyster and Neill, 1968), as shown in Fig. 2. In the epithermal spectrum, strong flux depressions occur at the resonance energies of ^{238}U (Johansson, 1968) (see Fig. 3). The shape and the width of these resonances depend on the type of moderator, the ratio of fuel to moderator, and on the geometry of the irradiation position. For elements showing resonances in the thermal and low epithermal range, activation depends on the temperature of the moderator. Examples taken from Volpe (1968) are given in Table I. Fluctuations in the neutron flux can cause further uncertainties with isotopes having half-lives considerably shorter than the time of irradiation.

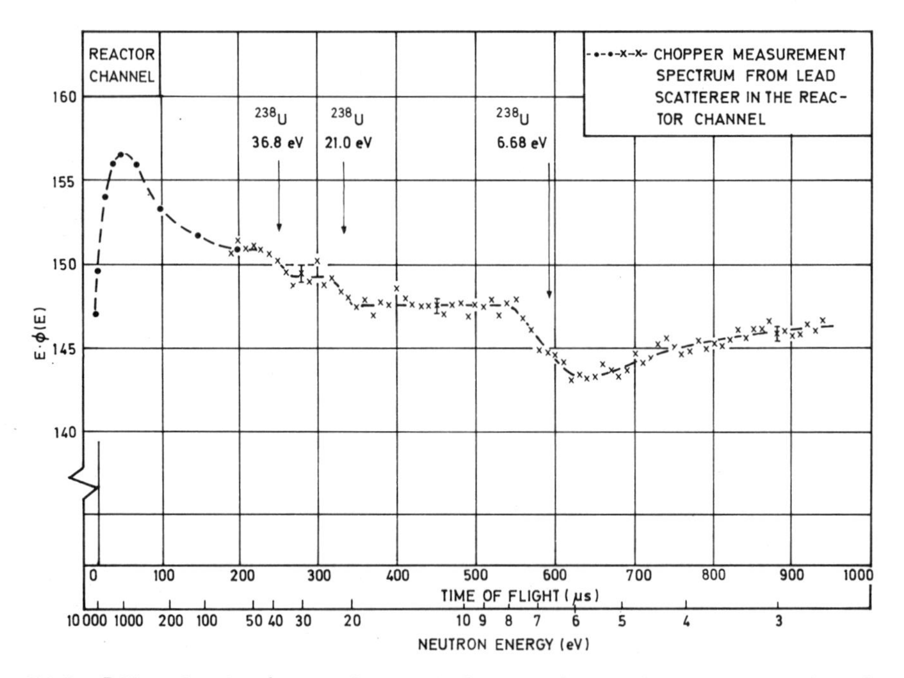

FIG.3. Epithermal spectrum from a lead scatterer in the reactor channel. Similar results were obtained from a graphite scatterer. The results are normalized to those in Fig.2. The errors shown are due to counting statistics only (after Johansson, 1968).

TABLE I. THERMAL COLUMN SPECTRUM NORMALIZATION (VOLPE, 1968)

Computation method	Temp. (K)	^{164}Dy	^{176}Lu	^{151}Eu
Pure Maxwellian	293.6	1.000	1.397	0.7923
	320	1.000	1.554	0.7711
	350	1.000	1.739	0.7493
	380	1.000	1.924	0.7299
	400	1.000	2.046	0.7180

Single-element standards

Single-element standards can be used for any element, but with geochemical materials in which many elements have to be determined in one sample, twenty to thirty standards have to be prepared. This is both tedious and time consuming. Solutions of two or three suitable elements can reduce the necessary work.

Standards for single elements are mostly prepared from solutions. In our laboratory, stoichiometric oxides of highest available purity (99.99+%) are dissolved in the required amount of very pure acid. The concentrations of metals in the solutions are close to, but always above, 1 mg of metal per gram of solution.

The solutions are prepared on a weight basis and are also measured by weight when used as standards. For single investigations, 10 μl of the standard solution are taken up in the tip of a micropipette and weighed on a sensitive analytical balance to an accuracy of 0.001 mg. The solution is then expelled and the emptied tip weighed in order to obtain the exact amount of solution used. The standard solution is evaporated on a piece of filter paper a few square millimeters in size. After evaporation the paper is folded and placed in an envelope made of 99.9999% pure aluminium foil 30 μm thick. Owing to the purity of the aluminium the measurement can be performed later without removing the material.

Standards prepared in this way introduce much less than 1% errors into the determinations. Some standard solutions are subject to alterations due to precipitation and (or) adsorption. This necessitates frequent checks.

Geochemical reference materials

Geochemical reference materials have been distributed by the US Geological Survey and other institutions. Flanagan (1974) gave a complete list of distributors of reference samples. Nevertheless, the true concentrations of elements are not known precisely. There are many gaps in our knowledge of the concentrations of several elements in basic and ultrabasic rocks because these elements are present at concentrations below the detection limit of most analytical techniques. The published values vary considerably. These variations may be due to (1) inhomogeneity in the reference sample materials, and (2) systematic differences between the analytical techniques.

An accurate analytical value can only be obtained if the standards are correct and if they provide all the necessary information for exact determinations.

Thus, rock references must have a satisfactory homogeneity so that a sample of 50 - 200 mg with a grain size below 30 μm provides the necessary data for standardization. Apart from the excitations of the particular isotopes in the geochemical standards, additional excitations occur. This causes serious interferences so that an undisturbed reference value for determining concentrations of elements in unknown samples can rarely be obtained. Mostly, the concentrations of elements in the unknown sample can be calculated from values in the geochemical reference samples, but with large errors. Table II gives examples of elements that cause serious interference. Because of the complex spectra obtained from geological samples, the general and exclusive use of geochemical reference samples as standards cannot be recommended.

Without exception the concentrations of those isotopes in irradiated geochemical reference samples can be used for making determinations in other materials which have interference-free lines (mostly of higher energy). The elements that can be considered are: Na, Sc, Fe, La, Hf. Whether their exact concentration is known and whether the homogeneity

TABLE II. EXAMPLES OF INTERFERENCES IN GAMMA SPECTRA OF ROCKS

Tb 933.96 keV	Eu 934 keV
Ce 143 keV	Fe 145 keV
Sb 603 keV	Cs 604 keV
Hg 279 keV	Se 279 keV
Zn 1115 keV	Sc 1120 keV
	Ta 1115 keV
	Eu 1115 keV

TABLE III. SUITABLE IRRADIATION PARAMETERS FOR GEOCHEMICAL SAMPLES
(F.R. 2 Reactor, Karlsruhe, F.R.G.)

Irradiation time	1 min	1 min	1 h	1 d	1 d
Decay time	1 h	1 d	1 d	8 d	30 d
Neutrons	Thermal	Thermal	Epithermal	Thermal	Thermal
Elements	Dy Dy-165	Sm Sm-153	Ca Sc-47	Na Na-24	Sc Sc-46
	Mn Mn-56	K K-42	Ni Co-58	As As-76	Cr Cr-51
			Ga Ga-72	Br Br-82	Fe Fe-59
			As As-76	Rb Rb-86	Co Co-60
			Sm Sm-153	Te I-131	Zn Zn-65
			Gd Gd-159	Ba Ba-131	Se Se-75
			Ho Ho-166	La La-140	Ag Ag-110m
			U Np-239	Nd Nd-147	Sb Sb-124
				Lu Lu-177	Cs Cs-134
				U Np-239	Ce Ce-141
					Eu Eu-152
					Tb Tb-160
					Yb Yb-169
					Hf Hf-181
					Ta Ta-182
					Hg Hg-203
					Th Pa-233

is sufficient are still open questions. With further refinement in measuring, the elements Co, Sb, Rb, Cs, Nd and Eu can be determined as well.

The limitation in the application of geochemical reference materials as standards does not exclude their use as test samples which should be run together with each series of unknowns in order to monitor the preparation of the single-element standard.

Conditions of irradiation

To determine a large number of elements by instrumental neutron activation analysis (INAA) the sample has to be irradiated several times. The scheme followed in our Institute for the different cooling times before measurement, the conditions of analysis and the elements which can be determined by a particular irradiation are given in Table III.

Additional considerations are necessary for investigations with epithermal neutrons. Because the resonance integrals of the isotopes to be activated are not known as exactly as the thermal cross-sections, the use of flux monitors is restricted. In this case only single-element standards can be used which should be applied in amounts similar to the concentrations of the samples. According to our experience epithermal irradiation can be applied successfully to those rocks in which thermal neutrons generate a high activity of, for example, Na and Fe. The favourable resonance integrals could be another reason for good determination. In our Institute Ca, Ni, Ga, As, (Sm)Gd, Ho and U are determined in rock samples after epithermal irradiation and comparison with single-element standards.

CONCLUSIONS

The accuracy of instrumental activation analysis is governed by many factors which influence each other. In geochemistry this accuracy depends on:

1. The accuracy of the standards used,
2. The homogeneity of the neutron flux in the total volume in which sample and standard are irradiated,
3. The homogeneity of the unknown sample,
4. The total spectrum generated in the sample (line interferences),
5. The type (resolution etc.) of measuring equipment,
6. The measuring time devoted to the sample, and
7. The evaluation procedure.

The importance of geochemical reference samples remaining for checking out analytical methods and single-element standards has been mentioned already.

To augment the rock references in circulation, the Institut für Petrographie and Geochemie der Universität Karlsruhe has prepared the following five new geochemical reference materials:

Hornblende	from	Norway
Phonolith	from	Kaiserstuhl, Upper Rhine Valley
Andesite	from	Santorini (Greece)
Latiandesite	from	Santorini (Greece)
Granite	from	Albtal, Black Forest

Some of these reference materials contain high concentrations of rare-earth elements. About 50 kg of each material has been prepared and made up into 25-g samples for distribution. Preliminary neutron activation analyses of these samples have been made and when sufficient data are available for a statistical evaluation, the initial results will be published.

REFERENCES

BEYSTER, J.R., NEILL, J.M. (1968), "Status of thermal neutron spectra: a review", Neutron Thermalization and Reactor Spectra (Proc. Symp. Ann Arbor, 1967) 2, IAEA, Vienna, p.3.

DAY, D.H. (1968), "Neutron spectra near to a temperature discontinuity in graphite", Neutron Thermalization and Reactor Spectra (Proc. Symp. Ann Arbour, 1967) 2, IAEA, Vienna, p.111.

FLANAGAN, F.J. (1969), U.S. Geological survey standards — II. First compilation of data for the new U.S.G.S. rocks, Geochim. Cosmochim. Acta 33 p.81.

FLANAGAN, F.J. (1973), 1972 Values for international geochemical reference samples, Geochim. Cosmichim. Acta 37 p.1189.

FLANAGAN, F.J. (1974), Reference samples for the earth sciences, Geochim. Cosmochim. Acta 38 p.1731.

INTERNATIONAL ATOMIC ENERGY AGENCY (1974), Handbook on Nuclear Activation Cross-Sections, Technical Reports Series No.156, Vienna.

JOHANSSON, E. (1968), "Thermal and epithermal neutron spectra in heterogeneous systems", Neutron Thermalization and Reactor Spectra (Proc. Symp. Ann Arbor, 1967) 2, IAEA, Vienna, p.77.

KIM, J.I., STÄRK, H. (1971), "Study on the monostandard activation analysis and its application to geological samples: investigation of Scheelite Deposits in the east Alps", Activation Analysis in Geochemistry and Cosmochemistry, Proc. Nato Advanced Study Institute, Kjeller, Norway, 7-12 Sep. 1970, Universitetsforlaget, Oslo, p.397.

LEDERER, C.M., HOLLANDER, J.M., PERLMAN, I. (1967), Table of Isotopes, 6th edn, John Wiley & Sons, Inc., New York.

RANDLE, K. (1974), Some trace element data and their interpretation for several new reference samples obtained by neutron activation analysis, Chem. Geol. 13 p.237.

SCHOCK, H.H. (1973), Instrumentelle Neutronenaktivierungsanalyse von Gesteinen am Beispiel von 7 Standardgesteinen des US Geological Survey, Z. Anal. Chem. 263 p.100.

SEELMANN-EGGEBERT, W., PFENNIG, G., MÜNZEL, H. (1974), Nuklidkarte, 4th edn, Bundesminister für Forschung und Technologie, Bonn.

STEINNES, E. (1971), "Epithermal neutron activation analysis of geological material", Activation Analysis in Geochemistry and Cosmochemistry, Proc. Nato Advanced Study Institute, Kjeller, Norway, 7-12 Sep. 1970, Universitetsforlaget, Oslo, p.113.

Van der LINDEN, R., DE CORTE, F., HOSTE, J. (1974), A compilation of infinite dilution resonance integrals, II., J. Radioanal. Chem. 20 p.695.

Van der LINDEN, R., DE CORTE, F., HOSTE, J. (1974), Activation analysis of geological material using ruthenium as a multiisotopic comparator, J. Radioanal. Chem. 20 p.729.

VOLPE, J.J. (1968), "Comparison of thermal neutron spectrum measurements and calculations in light-water-moderated uranium cells", Neutron Thermalization and Reactor Spectra (Proc. Symp. Ann Arbor, 1967) 2, IAEA, Vienna, p.209.

SUMMARY OF THE MEETING

SUMMARY OF THE MEETING

1. INTRODUCTION

The principal objectives of the Panel were to discuss the current status and development of nuclear techniques in the exploration, extraction and processing of mineral resources. From this comprehensive review of the current state of the art, conclusions can be made with respect to appropriate techniques to recommend for similar applications in the developing countries.

The participants in the Panel reviews and discussions were from eleven countries and had expertise and experience in most fields of applied nuclear techniques. Working papers were presented in six broad areas of technology as a basis for the discussions.

The Panel identified techniques which it considered useful in various stages of exploration and exploitation of mineral resources and hence worth recommending for use in developing countries. A summary of the discussions and conclusions of the Panel is presented here.

2. OIL, GAS AND COAL

The Panel reviewed the use of nuclear techniques, including stable and natural radioactive isotopes, in exploration and production of energy resources. Carbon isotope ratios in natural gas and crude oil were shown to have important geochemical and, ultimately, economic significance. Applications occur in the correlation of crude oils, the identification of source rocks and their correlation with crude oils, and in the determination of the maturity of source rocks by isotopic analysis of the natural gases from these rocks. Nuclear logging of boreholes is in world-wide use for locating and evaluating commercial deposits of hydrocarbons, coal, uranium and other useful minerals. The principal uses of nuclear logs are:

(1) for qualitative geological recognition of formations and correlation of wells,
(2) for identification of hydrocarbons, coal, uranium and oil shale,
(3) for quantitative evaluation of porosity, bulk density, uranium ore grade, and shale content.

The long spaced, dual detector neutron log is especially used to measure porosity and, in combination with a scattered gamma-ray density log, to identify gas. Small-diameter borehole accelerators are used as sources of pulsed 14-MeV neutrons to make routinely neutron die-away logs to distinguish hydrocarbons from salt water in cased holes. New accelerator logs based on spectral gamma-ray measurements are being developed to distinguish hydrocarbons from fresh water and to improve lithology determinations. Natural gamma-ray spectral measurements are used to solve correlation problems and to identify uranium-, thorium- and potassium-rich minerals. In coal exploration natural gamma-ray and scattered gamma-ray density logs are used to locate the coal and, in favourable circumstances, to estimate the ash content. For oil shale evaluation the density log is being used in small-diameter holes in Colorado oil shale to determine the potential oil yield. A new method of uranium detection based on the measurement of delayed fission neutrons produced by bombarding the formations with 14-MeV neutrons from a pulsed accelerator source has recently been described; within about one year this technique should be available as a commercial service in the USA and Canada. The use of californium-252 for this purpose is considered much less practical because of the radiation hazard in handling this neutron source.

For all these techniques increased attention needs to be given to the statistical significance of the data and their interpretation for geophysically and geologically significant information. The present techniques applied in geological exploration for mineral resources require proper evaluation of the data obtained from different geological, geophysical and geochemical methods (both nuclear and non-nuclear). One strives to obtain information about the geological formations being considered at the highest confidence level. This can be achieved by using a statistical approach to the problem. Special attention should be given to the influence of heterogeneity of geological formations on the available data.

Attention was called to four new books [1-4] published in the USSR covering the application of new techniques in geology, geochemistry and petroleum exploration. These works are especially important because there has been no new text in a western language covering these topics in about ten years. The new books have broad coverage and fill a current need in the scientific literature.

Tracer techniques have been developed and are currently being used in a number of applications in the drilling, completion and treatment of oil wells and in secondary recovery operations [5, 6]. Such services are available from specialized organizations in several countries. In mining operations, tracers are occasionally used for studying problems concerning ventilation and mine water.

TABLE I. NUCLEAR WELL LOGS

Objective	Logs used
Well-to-well correlation	Gamma ray, neutron
Fluid-filled porosity	Neutron, gamma ray
Hydrocarbon identification	
a. Oil-cased hole	Neutron die away
b. Gas-open hole	Neutron, gamma ray
Lithology	Neutron Spectral gamma ray (natural and induced)
Coal	
a. Identification	Gamma ray
b. Ash content	Gamma ray
Uranium	Gamma ray, neutron
Oil shale	Gamma ray

Tables I and II summarize the well logging methods routinely used in the petroleum industry. Table I shows only nuclear logs whereas Table II includes nuclear, electrical and acoustic (sonic) logs, thereby illustrating the inter-relationship between nuclear and other techniques in practical industrial applications.

In exploration for oil and gas, airborne techniques are used for regional reconnaissance surveys. The only nuclear technique used is airborne radiometric surveying to outline surface geology. The primary airborne technique in common use is aeromagnetic surveying.

3. METALLIC AND NON-METALLIC MINERALS

3.1. Exploration

3.1.1. Airborne surveys

3.1.1.1. Radiometric surveys

This subject has been considered in several previous publications of the IAEA [7, 8, 9] and the subgroup has been concerned with only the most recent progress made in this area. Particular consideration was given to:

(i) radiation field modelling
(ii) predicting the interaction of gamma radiation with NaI(Tl) crystals
(iii) predicting system performance

It was concluded that more attention should be given to the data on the geometric cross-section effects of the detector which are available in the USAEC, Health and Safety Report [10] and are particularly relevant to airborne applications. From these data, potential users are able to assess the optimum size of crystals for particular applications. Data of a similar kind are also now becoming available from the Risø Laboratory of the Danish Atomic Energy Commission [11]. It is stressed that detectors used for such applications should be sufficiently large, otherwise the statistical significance of collected data will be inadequate.

It is now usual to display the data "on-line" in an analogue form, but it is highly desirable that all data should be collected on tape in a digital form for later processing at base. It is also recommended that tapes be "played back" in the field so that any deficiencies in the data can be identified and, if necessary, rectified by re-flying the defective zones before the aircraft is returned to base.

Calibration platforms are an expensive construction but they are essential for allowing intercomparisons of various spectrometer systems. It is therefore recommended that the Agency consider the possibility of promoting international co-operation by providing such facilities at selected sites.

It is desirable that potential users of these facilities should clearly understand whether they want to do simple reconnaissance or detailed spectrometric surveys. The former is substantially cheaper, hence the need for justifying the greater expense of the more complicated spectrometric technique.

TABLE II. TYPES OF LOGS FOR LITHOLOGICAL AND POROSITY DETERMINATION IN THE OIL AND GAS INDUSTRIES

Mud and hole conditions	Data desired	Formation type		
		Unconsolidated formations (high porosity)	Medium formations (15 - 25% porosity)	Hard formations (low porosity)
Fresh muds (water base)	Lithology	Induction-electrical survey Electrical survey Neutron Gamma ray	Induction-electrical survey Electrical survey Neutron Gamma ray Sonic	Induction-electrical survey Laterolog Electrical survey Natural gamma ray Neutron Gamma ray Sonic
	Porosity	Sonic Gamma ray Neutron Microlog	Sonic Gamma ray Microlog Neutron	Sonic Neutron Gamma ray Microlaterolog
Salt muds (water base 20 000 ppm chlorides)	Lithology	Electrical survey Induction-electrical survey Laterolog Natural gamma ray Sonic Epithermal neutron Gamma ray	Laterolog and natural gamma ray Electrical survey and natural gamma ray Gamma ray Sonic Epithermal neutron	Laterolog and natural gamma ray Gamma ray Sonic Epithermal neutron
	Porosity	Sonic Gamma ray Microlog Epithermal neutron	Sonic Gamma ray Microlaterolog Epithermal neutron	Sonic Epithermal neutron Gamma ray Microlaterolog
Oil-base muds	Lithology	Induction log, natural gamma ray, gamma ray, sonic, neutron		
	Porosity	Sonic, gamma ray, neutron		
Empty or gas-filled hole	Lithology	Induction log, natural gamma ray, density		
	Porosity	Gamma ray and neutron		
Cased hole	Lithology	Natural gamma ray, neutron, sonic		
	Porosity	Neutron and sonic		

The general conclusion is that an increased use has been made of airborne surveys with processing mostly being done at ground level. There is also an increasing tendency to use four-channel spectrometry now that improved equipment is available.

The value of the radiometric method for uranium surveying is clear and need not be further emphasized here. The value of spectrometry in geological mapping based on the differences in U, Th and K contents of different rock types is also clear. Other areas of usefulness are:

- bauxite (which is often associated with thorium and some uranium [12])
- phosphates (which are sometimes associated with uranium)
- certain beach sands (which contain titanium, zirconium and rare earth minerals which, in turn, contain either uranium or thorium or both these elements)
- potassium mineral distributions as a guide to base metal concentrations, e.g. porphyry copper
- measurement of the depth of snow cover for flood control and estimation of water availability for hydroelectric power generation.

Final note: In radiometric surveys care must be taken to interpret the data properly and to avoid confusing overburden data with bedrock data.

3.1.1.2. Surveys of airborne particulates and gases

The value of making surveys of airborne particulates and gases is yet to be established as the "tracer" may be far removed from its origin. Nuclear techniques have an application, but only in the analysis of the collected material and probably then only if detection limits of simpler techniques are exceeded and more sensitive techniques are required.

3.1.2. Land-based surveys

3.1.2.1. Car-borne techniques

The concentrations of U, K and Th can be mapped by means of vehicle-mounted systems. Though these systems cannot cover such large areas of territory as those covered by an aeroplane, they can be used to explore localized areas in detail. Many of the remarks made about airborne spectrometers apply equally to vehicle-mounted and portable systems, but there are also some problems that are specific to the vehicle and portable systems.

3.1.2.2. Use of field laboratories

(i) Analysis of solid samples

Materials collected from outcrops of ore bodies, stream sediments, drill cores, and drill chippings from percussion holes are analysed for geochemical or mine-control purposes. Portable XRF equipment using balanced filters is in widespread routine use for single- and four-element analysis in the field, mainly for mine control. However, the limits of detection are not usually low enough for geochemical analysis. Portable equipment based on solid-state detectors is being evaluated for high-sensitivity field analysis.

EDXRF equipped with solid-state detectors has detection limits of a few ppm. It has an inherent multi-element capability, does not usually require sample preparation and is non-destructive. The technique is therefore under active evaluation and is beginning to be used routinely, especially for geochemical analyses. At present, however, atomic absorption spectrometry is still the most widely used method for trace and minor element analyses of collected samples.

Whether it is desirable to take samples back to the laboratory for analysis or to do such analysis in the field is a matter of policy. Such policy may be decided as much by relative cost as by the wishes of the supervising geologist.

If samples are removed to a principal laboratory, nuclear techniques can be applied for analyses. In particular, neutron techniques (principally neutron activation and gamma-ray excitation), but these would require access to reactors or to accelerators and are therefore non-competitive with EDXRF unless some special problems arise with the latter, such as inadequate sensitivity.

Therefore it is desirable to have tables of sensitivities readily available in, for instance, a guide book to nuclear applications. Such a guide book could contain the activation and gamma-ray excitation sensitivities for all the elements as well as the EDXRF sensitivities, so that the reader can judge when and if it is necessary to resort to the more complex and costly activation and gamma-excitation techniques.

It should also be noted that gamma-ray spectrometry is applicable to scanning and to analysis of drill cores: equipment may be automated.

(ii) Analysis of water samples

Atomic absorption spectrometry is generally applicable with or without pre-concentration, but for $pp10^9$ levels other methods are needed, e.g. for uranium, delayed neutron analysis using high thermal neutron fluxes is preferred. For other elements neutron activation is applicable and EDXRF is now under development. For radon in water, alpha counting is applicable [13].

(iii) Analysis of plant material

The value of this technique has not been fully established. It is likely that sensitive methods of analysis would be needed to establish levels of elemental concentrations and hence to deduce any correlation between plant occurrence and specific element distribution. Nuclear analytical techniques can be used for this purpose.

(iv) Analysis of ground gases

To measure ground radon concentrations as a means of deducing uranium anomalies, alpha counting is applicable as for the analysis of water. Analysis of other gases (e.g. He, SO_2, H_2S) is being considered to give information on uranium and sulphide ore bodies, but it is not envisaged that nuclear techniques will be widely used for these applications.

3.1.2.3. Borehole logging for exploration and mine control

Available techniques include:

The use of natural gamma radiation
The use of neutron activation
The use of neutron capture radiation
The use of neutron inelastic scattering
The use of gamma resonance radiation[1]
The use of selective gamma-gamma radiation(back-scattered radiation)
The use of neutron die-away technique.

It is necessary to distinguish here between exploration and mine development applications other than for lithological purposes for which there are in regular use techniques based on: natural gamma radiation (Fe, coal, Al as well as U, Th and K), gamma-ray back-scatter (density, porosity) (coal, uranium, iron ores); and neutron porosity methods.

Both natural-gamma and back-scatter methods can be made to yield quantitative data by correlation methods of iron ores and coal. Gamma-spectrometry methods are of course specific for U, Th and K and any element that can be suitably excited to give detectable radiation.

Instrumentation advances resulting from better data-handling techniques have assisted and will continue to assist developments, as also will the possibility of applying solid-state detector techniques. Stringent requirements on equipment stability are essential to obtain quantitative data. However, there are still some problems to be overcome before equipment based on Ge(Li) detectors can be used routinely. Neutron damage, microphony and the use of cryostats are the most important considerations. It is emphasized that strong support should be given to any serious attempts to develop high resolution solid-state detectors which do not require to be operated at low temperatures.

For quantitative work especially, there is a great need for stable, reliable neutron generator equipment capable of giving high neutron outputs which can be utilized to permit improved spectral resolution and allow time sequencing so that optimum use can be made of the more promising nuclear techniques.

It is important that, when suitable equipment does become available, the Agency support investigations to demonstrate its importance in mineral applications. This is very relevant to mineral resources development in the developing countries. It is known that several types of suitable neutron generator probes for borehole logging are in use in the USSR and it is hoped that these could be made available for use elsewhere. In the USA equipment has been in use for three years but it is not yet commercially available for purchase. However, there are service companies which have this type of equipment for use on a service basis.

There is also a need for more quantitative work in this area to take into account the influence of such factors as water, hole diameter and rock type on the analytical results.

Several groups have published limited data on the quantitative estimation of elemental concentrations (e.g. Cu, Ni, Fe, Cr, Si, F, U) by neutron

[1] Only for large holes (diameter greater than 15 cm).

methods, but there is still a real need for more detailed publication of investigational work and of field data. Similarly, better organized physical data could be used in feasibility studies of nuclear analytical methods [14].

The Panel identified the important problems of calibrating borehole probes. Large experimental facilities are difficult to construct and chippings are used, but these are not fully satisfactory. A more accurate correlation can now be made between mathematical results obtained by using Monte-Carlo methods and those obtained with small model boreholes. This is a useful approach.

Statistical techniques are being developed to interpret these data and are expected to play an increasingly important role in the future.

3.1.3. Sea-bed survey of Continental Shelf and deep sea-bed

The most important nuclear techniques are based on the measurement of natural gamma radiation by means of a towed sea-bed spectrometer and neutron interaction techniques, although the latter are still under development. Applications include the geological mapping of the Continental Shelf and the detection of placer deposits. These methods can also be developed for deriving geotechnical properties of the sea-bed.

Deep-sea exploration for phosphorites has been done in the Red Sea and along the coast of India to detect monazite sands. Commercial recovery of manganese nodules is being developed by several groups [15], and all operations include EDXRF equipment on-board ship for analysis of the nodules.

3.2. Mine area developments

The most important applications are for mine developments and for the analysis of rock samples and borehole cores. Applications occur underground and in open-pit mines. Concentration limits relate to economic cut-off grade and are much higher than in exploration applications: requirements are well within the capability of nuclear techniques. EDXRF is finding applications in the analysis of chippings and cuttings, and XRF borehole logging for Sn, Cu and Mo is now in routine use. Gamma resonance scattering is being considered for use in blast holes in open-pit mines. Application is limited to Cu and Ni. Gamma-gamma probes are in routine use for measuring the thickness of coal left behind in the roof after mining. A method based on measuring the 4.43-MeV gamma ray from the $(n, n'\gamma)$ reaction on carbon is also being considered for this application. Natural gamma and gamma-gamma techniques have been developed to the stage where they could be applied to iron ores.

3.3. Processing

The measurements made on solids and fluids during processing are outlined in the following sub-sections. The use of nuclear techniques provides continuous and instantaneous measurement on-line for control of the plant process. The measurements are often made directly on-line or in stream, or on a sample by-line taken from the stream. Laboratory assay of samples, which is an alternative to in-stream analysis, is sometimes used for control purposes, but it is less preferable.

Tracer techniques have been of great value in providing a fundamental knowledge of the surface sorption phenomena underlying flotation and they are being used to investigate plant kinetics which have to be known for efficient control.

3.3.1. Nuclear techniques

3.3.1.1. X- and gamma rays

(a) fluorescence
(b) absorption
(c) scatter { Compton, resonance }
(d) Mössbauer
(e) activation

3.3.1.2. β-particles

(f) absorption
(g) scattering

3.3.1.3. Neutrons

(h) absorption
(i) moderation
(j) (n, γ) inelastic
(k) capture
(l) activation

3.3.1.4. Natural radioactivity

3.3.1.5. Tracers

(m) radioactive
(n) activable tracers
(o) stable isotopes

3.3.2. Applications

3.3.2.1. Introduction

The measurements made on fluids during processing are as follows (the technique is indicated in parenthesis):

Instruments

Density	: in pipes and processing vessels (b)
Level	: in hoppers and bunkers (b)
Mass flow	: on conveyors and in pipes (b)
Particle size	: slurries on-line (f)

Concentration	: solids and slurries in process streams (b) moisture of solids in hoppers and on conveyors (i) ash in coal (b, c) elements on-line and in the laboratory (3.3.1.1. - 3.3.1.4.) compounds in the laboratory (d)

Tracers

Flow velocity	: in pipes and vessels
Flow rate	: in pipes and open channels
Wear	: grinding balls
Adsorption	: surface phenomena underlying flotation processes
Residence time	: in process units
Transport dynamics	: characteristics of material transport, e.g. flotation circuits
Mixing characteristics	: of powders, slurries, fluids in process vessels.

3.3.2.2. Instruments

Density

Nucleonic density gauges are used extensively in mineral processing, particularly on ore slurries in pipes to control the feed rate. They are generally based on a gamma-ray transmission technique. The signal from the radiation detector is displayed directly in specific gravity units or solids content. The density gauge gives a measurement integrated over a large representative portion of the pipe cross-section, does not interfere with the flow, is unaffected by the abrasive nature of the product, and can be used for control.

These instruments have been available commercially for over 15 years and many thousands are installed. On a mineral processing plant up to 50 gauges may be used. Care has to be taken to keep a reasonably constant distribution of the solids across the pipe cross-section and to exclude air. Accurate calibration is not easy as it involves passing known concentrations through the gauge, but the gauges are most commonly required to give a precise indication for control rather than an accurate measurement.

Other density gauges based on a variety of techniques (gravimetric weighing of a section of pipe, magnetostriction, ultrasonics, etc.) are available, but the gamma-ray gauge is generally preferred for measuring slurries.

Level

Nucleonic level gauges are used to provide either an on-off signal or a continuous indication of variations in level for process control. Generally they are based on a gamma-ray transmission technique. Since they are mounted on the outside of the process vessels, they are unaffected by the bumps from lumps of ore, by corrosion or abrasion caused by the contained fluids and by environmental factors such as dust, wetness and extreme temperatures.

These gauges have been available for over 20 years and are now well developed. They are used extensively (many thousands throughout the world), particularly for difficult measurements such as those of ores in hoppers and of hydrocarbons in polymerisation plants. As many as several tens are used on a single plant.

Many other techniques are used for level measurement (capacitance, resistance, ultrasonics, microwaves, etc.) and the number of non-nucleonic level gauges in use far outnumber the nucleonic ones. However, for the really difficult measurements the nucleonic type is preferred.

Mass flow

Nucleonic gauges are used to measure mass flow of fluids in pipes and granular solids on conveyors. The mass flow of fluids in pipes is determined by means of a gamma-ray density gauge and a suitable flow-meter (usually electromagnetic). Granular solids on conveyors are measured by means of gamma-ray transmission to indicate mass per unit length and a tachometer or flow-meter to measure conveyor velocity. The signals from the transducers are combined and processed to indicate flow rate and total mass conveyed.

These gauges are now in extensive and routine use. Other mass flow-meters are available (based on other fluid density gauges and load cells) but when the advantages of the nucleonic type are important then their use is preferred.

Particle size

An on-line nucleonic device for determining particle size distribution of slurries is now commercially available and used to control milling and grinding operations. The slurry is passed through a helix and the concentration of the different sized fractions at different radii is measured at the outlet by a beta-transmission density gauge.

An ultrasonic transmission or reflection analyser for particle size has recently been introduced in the USA. Its potential seems greater than that of the nucleonic gauge.

Concentration

Gamma-ray absorption techniques are widely used throughout industry for determining the solids content of slurries. Moisture monitors using a radioisotope neutron source and detector in a back-scatter geometry are in routine plant use for measurements on solids in hoppers and on conveyors. The ash content of coal is determined in processing plants by X-ray back-scatter or absorption techniques, but the method is susceptible to significant error when applied to coals with extremely variable ash composition.

The concentration of the higher atomic number ($Z \geqslant 25$) elements in mineral slurries and in fluids is measured by radioisotope X-ray techniques. For mineral slurries, installations of commercial equipment have been made mainly in the last three years and techniques for the basic metal elements can be considered proven in practice. Measurements of concentration of elements, such as rare earths and silver, require the use of more sensitive and sophisticated equipment based on solid-state detectors, and only one installation is in routine plant use so far. X-ray techniques can be used for low Z elements in a few specific types of applications such as the widely used determination of sulphur in hydrocarbons and the determination of calcium in cement raw mix.

An alternative to measuring the concentration by radioisotope X-ray techniques is the more elaborate and expensive instrumentation based on wavelength dispersion.

More specialized techniques of determination of concentration depend on neutron and gamma-ray interactions. Their main potential is for low Z elements in mineral slurries and for most elements in material on conveyors and in hoppers (particle size severely restricts the use of X-ray techniques). On-stream installations are as yet relatively few in number. Many potential applications of considerable importance to industry are in the developmental stage.

Assay in the laboratory of samples taken from processing streams is being undertaken routinely by a wide variety of nuclear techniques. Radioisotope X-ray techniques are the most widely used.

3.3.2.3. Tracer techniques

Tracer techniques are well established and have been profitably used in a number of applications to mineral processing operations. In addition to tracer methods of general technical applicability, methods developed to meet specific problems encountered in the mineral processing technology have been utilized. A considerable number of tracer investigations carried out in this area concern flotation processes.

Tracer techniques can be expected to play an increasing role in studies of process dynamics in view of the recent development of more efficient evaluation techniques which directly integrate tracer methods into modern systems engineering. This use will be of significant economic importance through the achievement of more efficient process operation and control. It will also lead to a better exploitation of mineral resources because low-grade ore deposits can be extracted economically.

Specific applications

Flotation	: Fundamental studies of surface sorption phenomena underlying flotation processes. Plant investigations of process kinetics, especially of flow and macroscopic mixing as a means of obtaining sufficient information on the system characteristics for efficient process operation and control.
Ore sorting and crushing:	Determination of grain-size distributions of solids and the effect of grinding operations on particle size; fundamental studies of material breakage; measurement of wear of grinding balls in ball-mills.

General applications

Standard tracer techniques are in use for determining various quantities such as flow velocity and flow rate of fluids, dilution etc. and for leak testing and similar checking purposes.

4. STANDARDS AND INTERCOMPARISON

The papers presented describe standardization and inter-laboratory comparisons as applied to geochemical analysis of samples. Calibration and standardization of instruments used for in-situ analysis are not included. The types of analysis reported on are neutron activation, energy dispersive X-ray fluorescence (EDXRF) and radiometric techniques. Neutron activation and radiometric methods are widely used and accepted for geochemical analysis. However, the most widely used techniques for laboratory geochemical analysis are wavelength dispersive X-ray fluorescence and atomic absorption. EDXRF analysis of samples prepared as thin specimens is a relatively new and highly promising technique which, although established in the related field of trace-element analysis of solid pollutants, is only just beginning to be used in geochemical work.

Information on the IAEA programme of intercomparisons and range of Standard Reference Materials (SRM's) was also supplied. A number of other organizations throughout the world offer SRM's and conduct intercomparisons, but without much inter-organization liaison.

Two classes of standards are in use: natural materials such as soils and rocks that have been exhaustively analysed (SRM's), and artificial calibration standards containing weighed amounts of their components. The Panel considers that SRM's are inadequate for instrument calibration and are likely to remain so in the foreseeable future. The main reasons are: heterogeneity, giving rise to particle size and segregation errors; difficulty of obtaining good stability with time (especially for soils); the large number of standards needed, particularly in multielement analysis, compared with the time and effort required to prepare and certify the SRM; and the specific need for single-element standards.

Primary gravimetric calibration standards should comprise a single element or a few non-interfering elements, and should be prepared in a form compatible with the analysis method. Other artificial standards may be necessary. For example, single-component minerals such as pyrite can be used to calibrate the effects of particle size in EDXRF. More attention should be devoted to the development of artificial primary standards that have the previously mentioned characteristics.

Interlaboratory comparisons should continue so as to ensure that new laboratories and new techniques are adequately tested. Both SRM's and well-characterized artificial standards can be used in intercomparisons.

The requirements for standards for nuclear methods of trace analysis are at least as stringent as those for non-nuclear methods. Nuclear methods are very useful because of their ability to analyse a wide range of elements and a wide range of concentrations; they can also analyse the sample material directly.

5. FUTURE IAEA WORK

Specific recommendations for future work on the subject of nuclear techniques in geochemistry and geophysics were submitted to the Director General of the IAEA after the Panel meeting.

REFERENCES

[1] ALEXEYEV, F.A., et al., Application of Nuclear Methods in Oil and Gas Geology, Nedra, Moscow (1973).
[2] Nuclear Geology, Transactions of VNIIYaGG, Moscow (1974).
[3] Isotopes of Carbon in Oil and Gas Geology, Nedra, Moscow (1973).
[4] Radioactive and Stable Isotopes in Geology and Hydrogeology, Atomizdat, Moscow (1974).
[5] MOTT, W.E., DEMPSEY, J.C., "Review of radiotracer applications in geophysics in the United States of America", Radioisotope Tracers in Industry and Geophysics (Proc. Symp. Prague, 1966), IAEA, Vienna (1967) 111.
[6] ALEKSEEV, F.A., SREBRODOLSKY, D.M., "Radioactive tracers in geophysics", Radioisotope Tracers in Industry and Geophysics (Proc. Symp. Prague, 1966), IAEA, Vienna (1967) 133.
[7] INTERNATIONAL ATOMIC ENERGY AGENCY, Nuclear Techniques and Mineral Resources (Proc. Symp. Buenos Aires, 1968), IAEA, Vienna (1969).
[8] INTERNATIONAL ATOMIC ENERGY AGENCY, Uranium Exploration Methods (Proc. Panel Vienna, 1972), IAEA, Vienna (1973).
[9] INTERNATIONAL ATOMIC ENERGY AGENCY, Recommended Instrumentation for Uranium and Thorium Exploration, Technical Reports Series No. 158, IAEA, Vienna (1974).
[10] UNITED STATES ATOMIC ENERGY COMMISSION, Report HASL - 195, USAEC, Washington (1968).
[11] DANISH ATOMIC ENERGY COMMISSION, Risø Report No. 317 (in preparation).
[12] ADAMS, J.A.S., RICHARDSON, K.A., Econ. Geol. 55 (1960) 1653.
[13] INTERNATIONAL ATOMIC ENERGY AGENCY, Uranium Exploration Methods (Proc. Panel Vienna, 1972), IAEA, Vienna (1973).
[14] CLAYTON, C.G., SANDERS, L.G., "The use of nuclear data in the design of radiation instruments for mineral exploration and mining", Nuclear Data in Science and Technology (Proc. Symp. Paris, 1973) 2, IAEA, Vienna (1973) 391.
[15] Mining Magazine 131 3 (1974) 194.

LIST OF PARTICIPANTS AND OBSERVERS

Participants

CALDWELL, R.L.	Mobil Research and Development Corp., Research Dept., P.O. Box 900, Dallas, TX 75221, United States of America
CAMERON, J.F.	Nuclear Enterprises Ltd., Bath Road, Beenham, Reading RG7 5PR, United Kingdom
CLAYTON, C.G.	UKAEA, Atomic Energy Research Establishment, Mining Instrumentation, Building 7, Harwell OX11 ORA, United Kingdom
CZUBEK, J.A.	Institute of Nuclear Physics, ul. Radzikowskiego 152, 31-342 Cracow, Poland
LANDSTRÖM, O.	AB Atomenergi, Studsvik, Fack, S-611 01 Nyköping, Sweden
LJUNGGREN, K.	Isotope Techniques Laboratory, Drottning Kristinas väg 45-47, S-114 28 Stockholm, Sweden
LØVBORG, L.	Danish Atomic Energy Commission, Research Establishment Risø, DK-4000 Roskilde, Denmark
PRZEWŁOCKI, K.	Hydrology Section, Division of Research and Laboratories, International Atomic Energy Agency, Kärntner Ring 11, P.O. Box 590, A-1011 Vienna, Austria

PUCHELT, H.	Institut für Petrographie und Geochemie der Universität Karlsruhe, Kaiserstrasse 12, D-7500 Karlsruhe 1, Federal Republic of Germany
RHODES, J.R.	Columbia Scientific Industries, Applied Research Division, 3625 Bluestein Boulevard, P.O. Box 6190, Austin, TX 78762, United States of America
SHTAN', A.S.	USSR State Committee for the Utilization of Atomic Energy, Moscow, USSR
STAHL, W.	Bundesanstalt für Geowissenschaften und Rohstoffe, P.O. Box 23 0153, D-3000 Hannover, Federal Republic of Germany
WATT, J.S.	Australian Atomic Energy Commission, Research Establishment, New Illawarra Road, Lucas Heights, Private Mail Bag, Sutherland 2232, N.S.W., Australia
WYLIE, A.W.	CSIRO, Division of Mineral Physics, P.O. Box 124, Port Melbourne, Vic. 3207, Australia

Observers

BESWICK, C.K.	Nucleonic Data Systems Inc., European Office, Opernring 1/E/628, A-1010 Vienna, Austria
HECHT, F.	Analytisches Institut der Universität Wien, Währinger Strasse 38, A-1090 Vienna, Austria

POCHET, R.	Commissariat à l'énergie atomique, Département des prospections et recherches minières, Section géophysique, Bât. 78 - B.P. n°106, F-30200 Bagnols-sur-Cèze, France
SCHROLL, E.	Grundlageninstitut der Bundesversuchs- und Forschungsanstalt Arsenal, Obj. 210, A-1030 Vienna, Austria
SOMLYAI, Z.	Mecseki ércbánya vállalat, Pécs, Hungary
VADOS, I.	National Atomic Energy Commission, Országos Atomenergia Bizottság, Akadémia-utca 17, Budapest V, Hungary
WILKES, P.G.	Bureau of Mineral Resources, P.O. Box 378, Canberra, A.C.T. 2601, Australia

Secretariat

Scientific Secretary	FURUTA, T.	Division of Research and Laboratories, IAEA
Editor	ERICSON, Anne	Division of Publications, IAEA

HOW TO ORDER IAEA PUBLICATIONS

Exclusive sales agents for IAEA publications, to whom all orders and inquiries should be addressed, have been appointed in the following countries:

UNITED KINGDOM	Her Majesty's Stationery Office, P.O. Box 569, London SE 1 9NH
UNITED STATES OF AMERICA	UNIPUB, P.O. Box 433, Murray Hill Station, New York, N.Y. 10016

In the following countries IAEA publications may be purchased from the sales agents or booksellers listed or through your major local booksellers. Payment can be made in local currency or with UNESCO coupons.

ARGENTINA	Comisión Nacional de Energía Atómica, Avenida del Libertador 8250, Buenos Aires
AUSTRALIA	Hunter Publications, 58 A Gipps Street, Collingwood, Victoria 3066
BELGIUM	Service du Courrier de l'UNESCO, 112, Rue du Trône, B-1050 Brussels
CANADA	Information Canada, 171 Slater Street, Ottawa, Ont. K 1 A OS 9
C.S.S.R.	S.N.T.L., Spálená 51, CS-110 00 Prague Alfa, Publishers, Hurbanovo námestie 6, CS-800 00 Bratislava
FRANCE	Office International de Documentation et Librairie, 48, rue Gay-Lussac, F-75005 Paris
HUNGARY	Kultura, Hungarian Trading Company for Books and Newspapers, P.O. Box 149, H-1011 Budapest 62
INDIA	Oxford Book and Stationery Comp., 17, Park Street, Calcutta 16
ISRAEL	Heiliger and Co., 3, Nathan Strauss Str., Jerusalem
ITALY	Libreria Scientifica, Dott. de Biasio Lucio "aeiou", Via Meravigli 16, I-20123 Milan
JAPAN	Maruzen Company, Ltd., P.O.Box 5050, 100-31 Tokyo International
NETHERLANDS	Marinus Nijhoff N.V., Lange Voorhout 9-11, P.O. Box 269, The Hague
PAKISTAN	Mirza Book Agency, 65, The Mall, P.O.Box 729, Lahore-3
POLAND	Ars Polona, Centrala Handlu Zagranicznego, Krakowskie Przedmiescie 7, PL-00-068 Warsaw
ROMANIA	Cartimex, 3-5 13 Decembrie Street, P.O.Box 134-135, Bucarest
SOUTH AFRICA	Van Schaik's Bookstore, P.O.Box 724, Pretoria Universitas Books (Pty) Ltd., P.O.Box 1557, Pretoria
SPAIN	Nautrónica, S.A., Pérez Ayuso 16, Madrid-2
SWEDEN	C.E. Fritzes Kungl. Hovbokhandel, Fredsgatan 2, S-103 07 Stockholm
U.S.S.R.	Mezhdunarodnaya Kniga, Smolenskaya-Sennaya 32-34, Moscow G-200
YUGOSLAVIA	Jugoslovenska Knjiga, Terazije 27, YU-11000 Belgrade

---

Orders from countries where sales agents have not yet been appointed and requests for information should be addressed directly to:

Publishing Section,
International Atomic Energy Agency,
Kärntner Ring 11, P.O.Box 590, A-1011 Vienna, Austria

75-10196